I0468039

Potential Effects of Deepening the St. Johns River Navigation Channel on Saltwater Intrusion in the Surficial Aquifer System, Jacksonville, Florida

By Jason C. Bellino and Rick M. Spechler

Prepared in cooperation with the U.S. Army Corps of Engineers

Scientific Investigations Report 2013–5146

U.S. Department of the Interior
U.S. Geological Survey

U.S. Department of the Interior
SALLY JEWELL, Secretary

U.S. Geological Survey
Suzette M. Kimball, Acting Director

U.S. Geological Survey, Reston, Virginia: 2013

For more information on the USGS—the Federal source for science about the Earth, its natural and living resources, natural hazards, and the environment, visit http://www.usgs.gov or call 1–888–ASK–USGS.

For an overview of USGS information products, including maps, imagery, and publications, visit http://www.usgs.gov/pubprod

To order this and other USGS information products, visit http://store.usgs.gov

Suggested citation:
Bellino, J.C., and Spechler, R.M., 2013, Potential effects of deepening the St. Johns River navigation channel on saltwater intrusion in the surficial aquifer system, Jacksonville, Florida: U.S. Geological Survey Scientific Investigations Report 2013–5146, 34 p., http://pubs.usgs.gov/sir/2013/5146/.

Acknowledgments

The authors gratefully acknowledge the cooperation and assistance of Stephen Myers, Jeff Navaille, Steve Bratos, Steve Ross, and Jason Harrah of the U.S. Army Corps of Engineers, Jacksonville District.

Recognition is also given to U.S. Geological Survey (USGS) employees Joseph Hughes, W. Fred Falls, Howard Reeves, and Glen Carleton for their constructive technical review comments. Special thanks are given to Michael Deacon of the USGS Science Publishing Network (SPN) for his thorough editorial review.

Contents

Figures

Tables

Conversion Factors and Datums

Multiply	By	To obtain
Length		
inch (in.)	2.54	centimeter (cm)
inch per year (in/yr)	2.54	centimeter per year (cm/yr)
foot (ft)	0.3048	meter (m)
mile (mi)	1.609	kilometer (km)
Area		
square mile (mi^2)	2.590	square kilometer (km^2)
Flow rate		
cubic foot per second (ft^3/s)	0.02832	cubic meter per second (m^3/s)
gallon per day (gal/d)	0.04381	meter per day (m/d)
gallon per minute (gal/min)	0.06309	liter per second (L/s)
million gallons per day (Mgal/d)	0.04381	cubic meter per second (m^3/s)
inch per year (in/yr)	25.4	millimeter per year (mm/yr)
Hydraulic conductivity		
foot per day (ft/d)	0.3048	meter per day (m/d)
Hydraulic gradient		
foot per mile (ft/mi)	0.1894	meter per kilometer (m/km)
Transmissivity*		
foot squared per day (ft^2/d)	0.09290	meter squared per day (m^2/d)

Vertical coordinate information is referenced to the North American Vertical Datum of 1988 (NAVD 88).

Horizontal coordinate information is referenced to the North American Datum of 1983 (NAD 83).

Elevation, as used in this report, refers to distance above or below the vertical datum.

Transmissivity: The standard unit for transmissivity is cubic foot per day per square foot times foot of aquifer thickness [(ft^3/d)/ft^2]ft. In this report, the mathematically reduced form, foot squared per day (ft^2/d), is used for convenience.

Concentrations of chemical constituents in water are given either in parts per thousand (ppt) or milligrams per liter (mg/L).

Abbreviations

DOR	Department of Revenue
DP	deep observation point
EFDC	Environmental Fluid Dynamics Code
ET	evapotranspiration
FDOH	Florida Department of Health
GHB	general head boundary
L	length
lidar	light detection and ranging
M	mass
MLW	mean low water
MLLW	mean lower low water
NCDC	National Climatic Data Center
NED	National Elevation Dataset
NEXRAD	Next Generation Weather Radar system
NWIS	National Water Information System
NWS	National Weather Service
PPT	parts per thousand
RET	reference evapotranspiration
SH	shallow observation point
T	time
USACE	U.S. Army Corps of Engineers
USGS	U.S. Geological Survey

Potential Effects of Deepening the St. Johns River Navigation Channel on Saltwater Intrusion in the Surficial Aquifer System, Jacksonville, Florida

By Jason C. Bellino and Rick M. Spechler

Abstract

The U.S. Army Corps of Engineers (USACE) has proposed dredging a 13-mile reach of the St. Johns River navigation channel in Jacksonville, Florida, deepening it to depths between 50 and 54 feet below North American Vertical Datum of 1988. The dredging operation will remove about 10 feet of sediments from the surficial aquifer system, including limestone in some locations. The limestone unit, which is in the lowermost part of the surficial aquifer system, supplies water to domestic wells in the Jacksonville area. Because of density-driven hydrodynamics of the St. Johns River, saline water from the Atlantic Ocean travels upstream as a saltwater "wedge" along the bottom of the channel, where the limestone is most likely to be exposed by the proposed dredging. A study was conducted to determine the potential effects of navigation channel deepening in the St. Johns River on salinity in the adjacent surficial aquifer system. Simulations were performed with each of four cross-sectional, variable-density groundwater-flow models, developed using SEAWAT, to simulate hypothetical changes in salinity in the surficial aquifer system as a result of dredging. The cross-sectional models were designed to incorporate a range of hydrogeologic conceptualizations to estimate the effect of uncertainty in hydrogeologic properties. The cross-sectional models developed in this study do not necessarily simulate actual projected conditions; instead, the models were used to examine the potential effects of deepening the navigation channel on saltwater intrusion in the surficial aquifer system under a range of plausible hypothetical conditions.

Simulated results for modeled conditions indicate that dredging will have little to no effect on salinity variations in areas upstream of currently proposed dredging activities. Results also indicate little to no effect in any part of the surficial aquifer system along the cross section near River Mile 11 or in the water-table unit along the cross section near River Mile 8. Salinity increases of up to 4.0 parts per thousand (ppt) were indicated by the model incorporating hydrogeologic conceptualizations with both a semiconfining bed over the limestone unit and a preferential flow layer within the limestone along the cross section near River Mile 8. Simulated increases in salinity greater than 0.2 ppt in this area were generally limited to portions of the limestone unit within about 75 feet of the channel on the north side of the river.

The potential for saltwater to move from the river channel to the surficial aquifer system is limited, but may be present in areas where the head gradient from the aquifer to the river is small or negative and the salinity of the river is sufficient to induce density-driven advective flow into the aquifer. In some areas, simulated increases in salinity were exacerbated by the presence of laterally extensive semiconfining beds in combination with a high-conductivity preferential flow zone in the limestone unit of the surficial aquifer system and an upgradient source of saline water, such as beneath the salt marshes near Fanning Island. The volume of groundwater pumped in these areas is estimated to be low; therefore, saltwater intrusion will not substantially affect regional water supply, although users of the surficial aquifer system east of Dames Point along the northern shore of the river could be affected. Proposed dredging operations pose no risk to salinization of the Floridan aquifer system; in the study area, the intermediate confining unit ranges in thickness from more than 300 to about 500 feet and provides sufficient hydraulic separation between the surficial and Floridan aquifer systems.

Introduction

Jacksonville's main navigation channel lies within a 20-mile reach of the St. Johns River, extending from the Atlantic Ocean at Mayport to Commodore Point near downtown Jacksonville in Duval County, Florida (fig. 1). The channel supports commercial shipping activities and is also used by the U.S. Navy for operations at Naval Station Mayport.

The navigation channel, also known as Jacksonville Harbor, was initially dredged in 1896 when the U.S. Army Corps of Engineers (USACE) deepened the channel from 13.5 to 15.5 feet below North American Vertical Datum of 1988 (NAVD 88). The harbor was then dredged to a depth of 31–33 feet below NAVD 88 in 1910 and to 39–41 feet below NAVD 88 in 1977. Recent dredging to the current (2013) maintained depth of 41–43 feet below NAVD 88 was completed in two phases, the first of which was completed in 2003 and encompassed the section of the harbor from River Mile 0 to approximately River Mile 14. The second phase was completed in 2010, and deepened the section of the harbor from River Mile 14 to River Mile 20. To date, no studies have

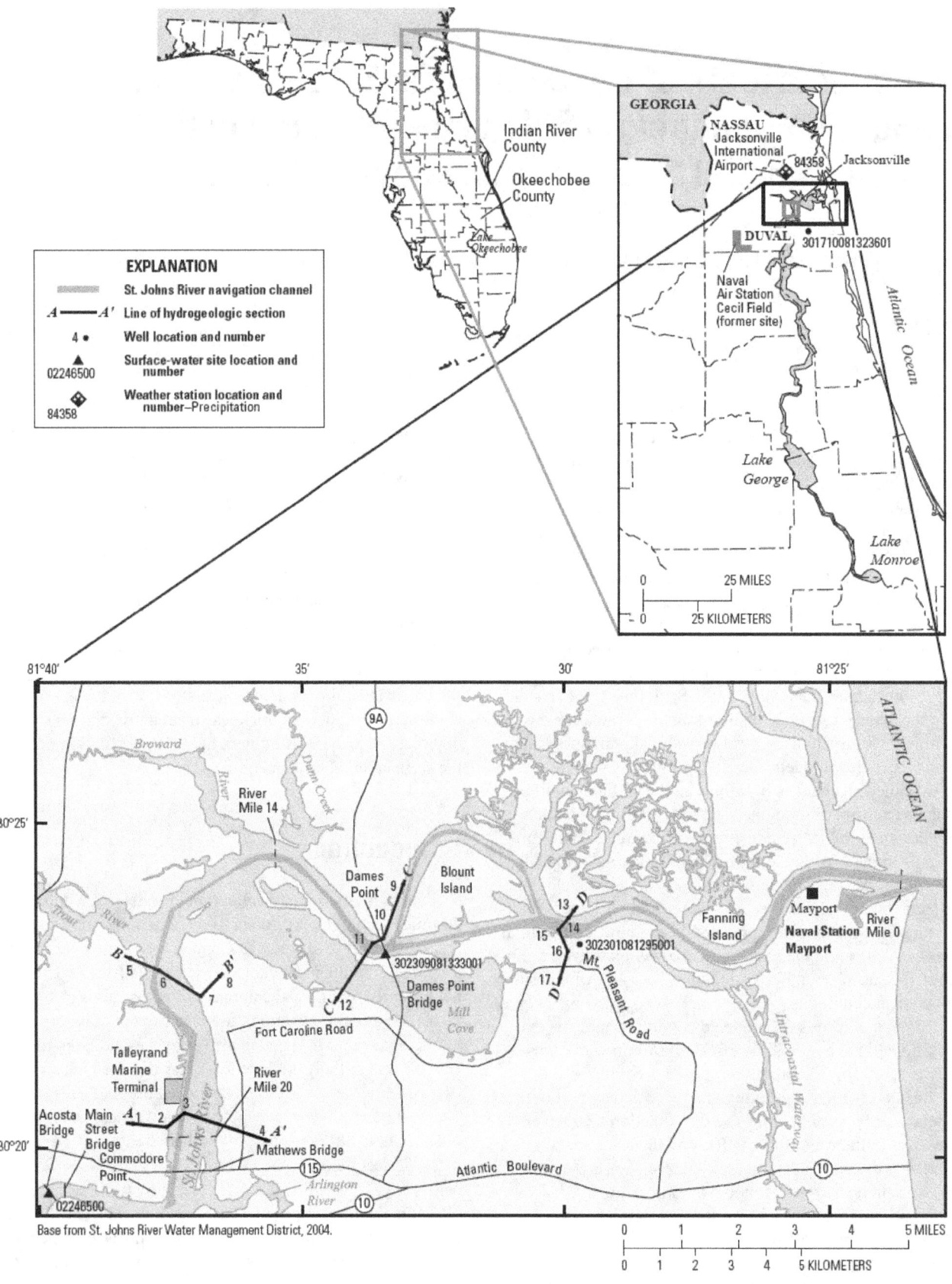

Figure 1.   Location of study area in Duval County, Florida.

been published that quantitatively evaluate the effects of these dredging activities on salinity in the surficial aquifer system.

The USACE and a nonfederal sponsor, the Jacksonville Port Authority, are evaluating the costs and benefits of a proposed USACE plan to further deepen a 13-mile reach of the navigation channel in the St. Johns River from its current depth to between 50 and 54 feet below NAVD 88. The U.S. Geological Survey (USGS), in cooperation with the USACE, initiated a study in 2010 to determine the potential effects of the proposed dredging on salinity in the surficial aquifer system adjacent to the St. Johns River. Variable-density groundwater-flow and salinity transport models were used to quantify the effects that deepening the St. Johns River navigation channel may have on saltwater intrusion in the surficial aquifer system and identify areas that may be susceptible to groundwater-quality degradation and, therefore, require future water-level and (or) water-quality monitoring.

Because other ports of the United States are expected to consider similar deepening plans, the approach developed for this study could be applied to other areas to evaluate the effects of channel deepening on saltwater intrusion.

Purpose and Scope

The purpose of this report is to document how the proposed deepening of the St. Johns River navigation channel may affect salinity in the adjacent surficial aquifer system. Cross-sectional variable-density groundwater-flow and salinity transport models of the surficial aquifer system were developed along four sections that intersect the navigation channel. The models were developed to simulate the increase or decrease in aquifer salinity in response to channel deepening in the presence or absence of two hydrogeologic features: semiconfining beds and preferential flow paths. Model simulations were also conducted to determine the potential effects of a severe long-term drought through decreased water levels in the surficial aquifer system. The simulations were evaluated for the 121-day period from December 1, 1998, through March 31, 1999, to coincide with the available river stage and salinity results from the USACE Environmental Fluid Dynamics Code (EFDC) model and extended another 363 days with hypothetical river stage created from the 121 days of actual data. Several plausible hydrogeologic models were tested for their effects on the computed salinity distribution in the surficial aquifer system. The hydrologic and geologic data used to develop aquifer parameters and boundary conditions for the representative cross-sectional, variable-density numerical models are presented.

Previous Studies

Much of the geology of the study area has been described previously by Puri (1957), Puri and Vernon (1964), and Miller (1986). The groundwater resources of Duval County have been described by Leve (1966), Fairchild (1972), Causey and Phelps (1978), Phelps (1994), and Spechler (1994).

The physical and chemical characteristics of the St. Johns River are described extensively in published studies by the USACE, USGS, Florida Bureau of Geology, St. Johns River Water Management District, and private engineering firms. Examples include reports by Anderson and Goolsby (1973), which describes in detail the physical and chemical characteristics of the lower St. Johns River, and by Morris (1995), which describes the relations between water levels, velocity, flow, storage, and salinity in the lower St. Johns River and reviews previous hydrodynamic modeling studies.

Numerous studies have been conducted by the USACE that address navigational improvements to the Jacksonville Harbor. In addition, Spechler and Stone (1983) described the effects that deepening the navigation channel to 46–48 feet below NAVD 88 would have on the adjacent surficial aquifer system. The authors concluded that the dredging activities proposed at the time would not be expected to increase salinity in the surficial aquifer system.

Vertical Datums

The USACE typically uses a vertical datum that references tidal gages such that resulting elevations are in terms of feet above or below a given water level in a specified tidally affected area. This is particularly important when planning dredging activities for which the draft of large vessels must be taken into account to ensure the channel is navigable during low tides. By nature, these tidal datums do not specify a constant areal elevation, even at relatively local scales. The datum used by the USACE for the proposed dredging project is mean lower low water (MLLW), defined as "the average of the lower low water height of each tidal day observed over the National Tidal Datum Epoch" (National Oceanic and Atmospheric Administration, 2013). In the study area, MLLW is approximately equivalent to mean low water (MLW), the historical datum used by the USACE for the Jacksonville Harbor navigation project; differences between MLLW and MLW are about 0.1–0.2 foot, generally increasing from west to east. For convenience and to minimize confusion, MLLW was converted to NAVD 88 herein. The North American Vertical Datum of 1988 ranges from about 1 foot above MLLW near the Acosta Bridge to about 3 feet above MLLW at Mayport.

Description of Study Area

The study area is located in east-central Duval County, Florida, and covers an area of about 140 square miles (mi²), including the St. Johns River and adjacent areas from its mouth to about 24 miles upstream to Commodore Point near downtown Jacksonville (fig. 1). The topography within the study area is generally low and flat, with land surface elevations that range from sea level to about 85 feet above NAVD 88. The present topography of the study area is largely a consequence of marine terrace formation during global sea-level decline in the Pleistocene (Leve, 1966). North of the St. Johns River, land surface elevations generally do not exceed 30 feet. East of Blount Island, much of the area is covered by saltwater marshes, and elevations generally do not exceed 5 feet. South of the St. Johns River, elevations generally range from 30 to 50 feet, but are less than 20 feet near the Intracoastal Waterway.

The humid subtropical climate of Duval County is characterized by warm, relatively wet summers and mild, relatively dry winters. The mean annual rainfall during 1981–2010 was about 52 inches, measured at the Jacksonville International Airport northwest of the study area. Rainfall is unevenly distributed throughout the year; about half occurs during the wet season from June through September and the remainder occurs from October to May. Thunderstorms account for most of the summer rainfall, and tropical storms and hurricanes occasionally cause brief periods of widespread heavy rainfall and associated flooding.

Surface-water drainage is primarily through the St. Johns River and its tributaries. Principal tributaries of the river within the study area include the Broward River, Dunn Creek, Trout River, Arlington River, and the Intracoastal Waterway (fig. 1). The St. Johns River and its tributaries are tidal throughout the most of the study area.

St. Johns River

The St. Johns River is an elongated shallow river estuary and the longest waterway in Florida. From its headwaters in Indian River and Okeechobee Counties downstream to Jacksonville, the river generally flows south to north for a distance of more than 300 miles (Morris, 1995). At Jacksonville, in Duval County, the river turns eastward and flows another 15 miles until it discharges into the Atlantic Ocean at Mayport. Although the river is up to 3 miles wide, its width in the study area ranges from about 1,250 feet at the Main Street Bridge in Jacksonville to more than 2 miles at Mill Cove. The average water-surface gradient of the St. Johns River is approximately 0.1 foot per mile (ft/mi) (St. Johns River Water Management District, 1994, p. 24).

The river is subject to semidiurnal tides, and large volumes of seawater and freshwater are mixed during each tide-reversal cycle (U.S. Army Corps of Engineers, 1986, p. 73). Annual mean discharge for water years 1996–2012 (October 1, 1996, to September 30, 2012) at USGS station number 02246500 (St. Johns River at Jacksonville, FL) (fig. 1, table 1) was 7,345 cubic feet per second (ft³/s) (U.S. Geological Survey, 2013c).

Flow in the St. Johns River is modified considerably at times by wind and freshwater flow conditions. Northerly and northeasterly winds increase the velocity and duration of upstream tidal flows and decrease those of downstream tidal flows. Southerly and southwesterly winds have the opposite effect (Anderson and Goolsby, 1973, p. 15). Freshwater entering the river increases the volume and duration of downstream flows and decreases those of upstream flows (Anderson and Goolsby, 1973, p. 15).

Tidal fluctuations affect flows in the St. Johns River considerably, traveling 110 miles upstream to Lake George. On occasion, tides have been reported in Lake Monroe, 161 miles from the mouth (Morris, 1995, p. 12). The average tidal range at the mouth of the St. Johns River at Mayport is about 4.5 feet (St. Johns River Water Management District, 2008, p. 2–7). Further upstream, the amplitude of the tidal fluctuations decreases to about 3.2 feet at Dames Point and about 1.5 feet at the Acosta Bridge (St. Johns River Water Management District, 2008).

The chemical characteristics of water in the St. Johns River vary substantially over space and time and at any given instant reflect the opposing influences of freshwater inflow, from tributaries and groundwater discharge, and saltwater inflow from the Atlantic Ocean. The chemical composition of the river water varies from that of seawater near the mouth to freshwater farther upstream, reflecting the progressive mixing of seawater with fresher river water along the reach. The degree of mixing depends on the relative density of the water masses as well as the amount of local turbulence and the water

Table 1.   Location of groundwater and surface-water stations.

[USGS, U.S. Geological Survey]

USGS site identification number	Station name	Site type	Latitude	Longitude
02246500	St. Johns River at Jacksonville, FL	Stream	30°01'20"	81°39'56"
302309081333001	St. Johns River at Dames Point Bridge at Jacksonville, FL	Stream	30°23'09"	81°33'30"
302301081295001	DS-522 Fort Caroline National Memorial Park	Well	30°23'01"	81°29'38"
301710081323601	DS-520 St. Johns River Water Management District Observation Well at Jacksonville, FL	Well	30°17'10"	81°32'36"

velocity (Morris, 1995). A substantial vertical salinity gradient can develop in the water column of the river when freshwater inflows are relatively large. The relatively fresh, less dense layer of water can flow over a denser layer flowing in the opposite direction, with relatively little mixing at the interface (Morris, 1995). River salinity generally increases during periods of low freshwater flow, and decreases during periods of high freshwater flow (Anderson and Goolsby, 1973). Salinity concentrations at USGS station number 302309081333001 (St. Johns River at Dames Point Bridge at Jacksonville, FL) (fig. 1, table 1) ranged from about 0.3 to 38.7 parts per thousand (ppt) during 1996–2001 and 2003–2011 (U.S. Geological Survey, 2013a). Salinity concentrations at the Acosta Bridge ranged from 0.1 to 34.5 ppt during 1995–2001 (St. Johns River Water Management District, 2008).

Navigation Channel

Shipping is an important industry along the St. Johns River, and a range of vessels use the docks and terminals along the channel. Most of these vessels are bulk carriers, car carriers, tankers, and container ships. When the Panama Canal expansion project is completed in 2015, larger container ships are expected to begin visiting ports along the East Coast of the United States. These larger ships can be up to 1,400 feet long, 150 feet wide, and when fully loaded, draft as much as 50 feet (U.S. Army Corps of Engineers, 2009). To meet the needs of larger cargo ships projected to dock at Jacksonville Harbor in the future, the depth of the existing navigation channel must be increased from 41–43 feet to 50–54 feet below NAVD 88.

The navigation channel in the St. Johns River extends about 24 miles from the Atlantic Ocean to the Acosta Bridge in downtown Jacksonville and is maintained by the USACE (fig. 1). The main navigation channel, a 20-mile reach of the river, extends from the river mouth to the Jacksonville Port Authority Talleyrand Marine Terminal. From the entrance channel in the Atlantic Ocean to U.S. Naval Station Mayport (at about River Mile 0), the channel depth is about 43–45 feet below NAVD 88. The present main channel has an authorized depth of about 41–43 feet below NAVD 88 from River Mile 0 to River Mile 20, just upstream of the Talleyrand Marine Terminal (U.S. Army Corps of Engineers, 2011, p. 10). From River Mile 20 to Commodore Point (River Mile 22), the channel depth is 35–37 feet below NAVD 88. The channel depth is 31–33 feet below NAVD 88 from Commodore Point to the railroad bridge adjacent to the Acosta Bridge. Channel widths range from 400 to 1,200 feet along the entire 22-mile reach.

The depth of the currently proposed dredging activities will be the maximum dredging depth of the normal channel and includes 1 foot of allowable overdepth and 1 foot of required dredging depth as well as additional deepening in areas of advance maintenance in an attempt to reduce the frequency of maintenance dredging. Allowable overdepth is included to compensate for physical conditions at the time of dredging as well as inaccuracies in the dredging equipment itself; required dredging depth is included to ensure that the minimum design depth is met.

Land Use

Major land-use categories in the study area as of 2004 included residential, marsh or wetlands, open lands or forest, streams and waterways, commercial, industrial, recreational, and agricultural (fig. 2; St. Johns River Water Management District, 2004). Residential lands (composing 26 percent of the study area) are generally south and west of the St. Johns River. Residential land use includes residential—low or medium density (18 percent) and residential—high density (8 percent) classifications. Marsh or wetlands (22 percent) and streams and waterways (16 percent) are found throughout the study area. Open and forested lands (17 percent) are mostly concentrated in the northern part of the study area. Commercial land (10 percent) is generally scattered throughout the western and southern parts of the study area, whereas industrial land (6 percent) is concentrated along parts of the St. Johns River, mainly because of its use by the shipping industry. Recreational land (2 percent) is predominantly scattered throughout the southern half of the study area and along the coast. Agricultural land use (1 percent) is almost exclusive to the northern margin of the study area.

Water Use

Groundwater is the principal source of water in Duval County for public supply, commercial-industrial self-supplied, domestic self-supplied, agricultural irrigation, and recreation self-supplied use. The primary source of groundwater is the Floridan aquifer system, although about 5 percent is withdrawn from the surficial aquifer system (Richard L. Marella, U.S. Geological Survey, written commun., 2011). Most of the water withdrawn from the surficial aquifer system is obtained from wells tapping the limestone unit, often described locally as "rock wells." In 2005, groundwater withdrawals in Duval County totaled about 168.7 million gallons per day (Mgal/d) of which approximately 8.8 Mgal/d was withdrawn from the surficial aquifer system (Richard L. Marella, U.S. Geological Survey, written commun., 2011). These totals were modified slightly from those provided by Marella (2009, p. 11) to account for additional domestic well withdrawals for irrigation previously not considered. Of the six water-use categories, domestic self-supplied accounted for 96 percent of the total groundwater withdrawn from the surficial aquifer system in 2005. Withdrawals for the remaining categories included 2 percent for agricultural self-supplied and 2 percent for recreational self-supplied (Richard L. Marella, U.S. Geological Survey, written commun., 2011). Less than 0.01 percent was withdrawn for the public supply, commercial-industrial self-supplied, and power-generation categories, combined. Heat pumps were not included in the water-use computations because most heat pumps, both closed and open loop, consume little groundwater.

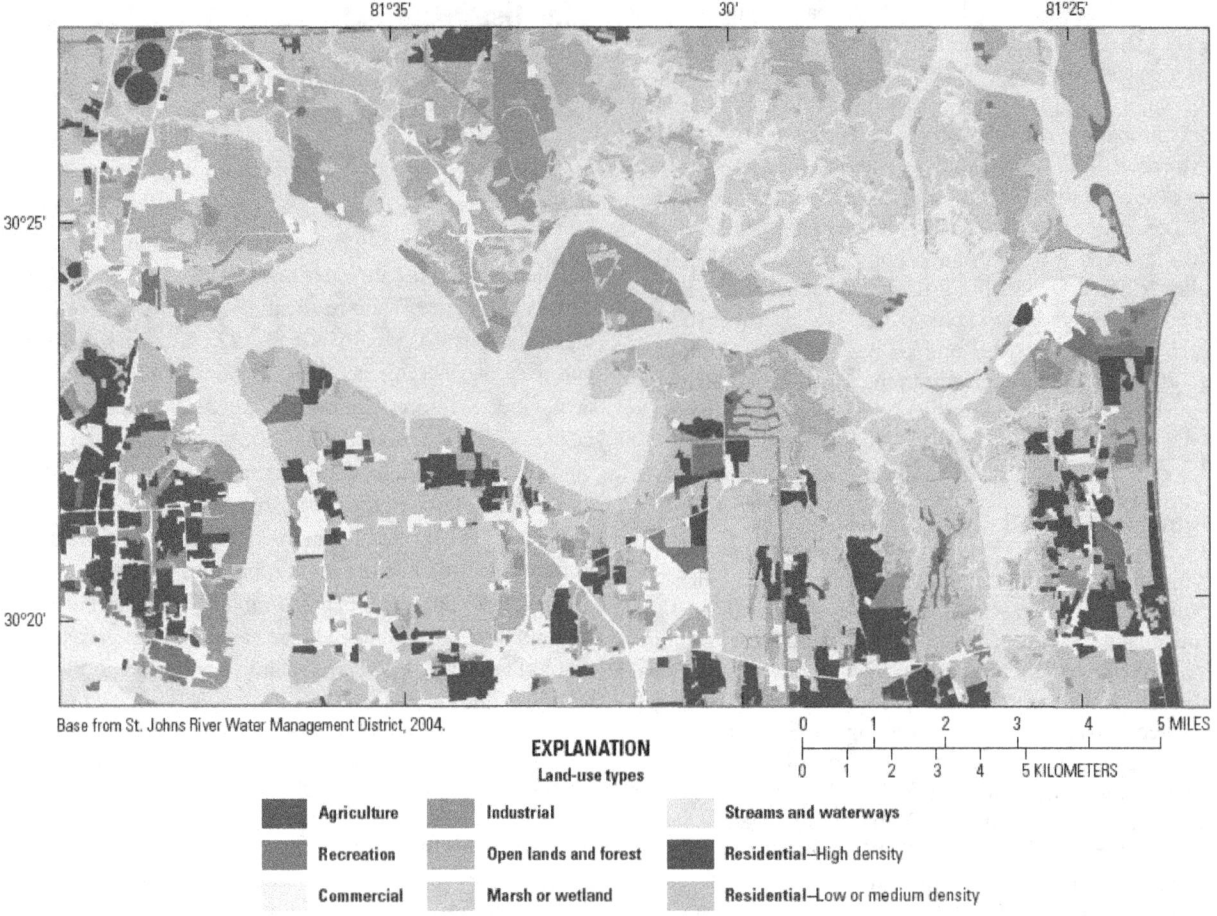

Figure 2.   Generalized land use in the study area, 2004, modified from St. Johns River Water Management District (2004).

Hydrogeology

The hydrogeologic system in eastern Duval County consists of a thick sequence of sedimentary rocks that include sands, clays, and carbonates. The three major hydrogeologic units present in the study area, in descending order, are the surficial aquifer system, intermediate confining unit, and Floridan aquifer system. The surficial aquifer system is the uppermost water-bearing unit and underlies all of Duval County. The intermediate confining unit, which consists of beds of relatively low permeability sediments that vary in thickness and areal extent, restricts the movement of water between the overlying surficial aquifer system and underlying Floridan aquifer system. Leakance estimates for the intermediate confining unit in Duval County and elsewhere in peninsular Florida range from 1×10^{-4} to 1×10^{-6} ft^{-1} (Tibbals, 1990; Boniol and others, 1993; Sepúlveda, 2002; Knochenmus, 2006). The lowermost water-bearing hydrogeologic unit underlying the county is the Floridan aquifer system, which is composed primarily of limestone and dolostone and is the principal source of water for Duval County (Marella, 2004, 2009).

Surficial Aquifer System

The surficial aquifer system underlies the entire study area and consists of interbedded lenses of sand, shell, clay, limestone, and dolostone. The sediments of the surficial aquifer system range from middle Miocene to Holocene age. The surficial aquifer system is divided into two water-bearing units, the water-table unit and the underlying limestone unit (fig. 3). These two units are separated by sediments of lower permeability that partially confine water in the limestone unit. The surficial aquifer system generally is 10 to 100 feet thick in the study area. The top of the intermediate confining unit is defined by the first occurrence of persistent beds of Miocene-age sediments, underlying the limestone unit, that contain a substantial increase in clay or silt. Variations in the thickness and dip of the hydrogeologic units based on geologic descriptions, core borings, and drillers' logs (locations shown in fig. 1) are depicted in four generalized hydrogeologic sections shown in figures 4 and 5.

Geologic age	Stratigraphic unit	Lithologic description	Hydrogeologic unit	
Holocene and Pleistocene	Undifferentiated surficial deposits	Sand, minor clay, and shell. Source of water to shallow sandpoint wells.	Surficial aquifer system	Water-table unit
– – – – – Pliocene or Late Miocene		Sand, clay, and some shell beds		Semiconfining bed
– – – – – Middle Miocene	Hawthorn Group	Interbedded limestone, sand, clay, and shell beds. Limestone is cavernous and sometimes dolomitic. Limestone is principal water-producing unit in the surficial aquifer system.		Limestone unit
		Clay, sand, limestone, dolostone, and phosphate	Intermediate confining unit	Confining bed

Figure 3. Generalized hydrogeology of the surficial aquifer system in the study area, modified from Fairchild (1972), Causey and Phelps (1978), Spechler and Stone (1983), and Scott (1988).

Water-Table Unit

The water-table unit composes the upper part of the surficial aquifer system; the unit is unconfined and consists of unconsolidated clastic deposits that range in age from Pleistocene to Holocene. The unit is composed primarily of fine- to medium-grained quartz sand that can contain thin beds of sandy clay. In places, especially near the coast, shell beds are present in these surficial sediments. The thickness of the water-table unit in the study area ranges from about 5 to 60 feet (figs. 4 and 5). The unit is not present in the navigation channel in most locations, having been removed by dredging.

The hydraulic properties of the water-table unit can vary considerably and are largely dependent upon aquifer thickness, physical characteristics (such as grain size and sorting), and the types of sediments that compose the aquifer. Few data are available concerning the hydraulic characteristics of the water-table unit in Duval County. Transmissivity values determined in an area 1.5 miles north of Blount Island averaged about 800 feet squared per day (ft²/d) (Spechler and Stone, 1983). At two sites in western Nassau County, transmissivity values ranged from about 100 to 950 ft²/d (Dames and Moore, 1987, p. AI–15). Horizontal hydraulic conductivities estimated from aquifer tests at the former Naval Air Station Cecil Field averaged 5 feet per day (ft/d) (Halford, 1998a, p. 25). Additional horizontal hydraulic conductivities determined from slug tests ranged from 0.6 to 5 ft/d (Halford, 1998a, p. 25). Horizontal hydraulic conductivities estimated from aquifer and slug tests at Naval Station Mayport ranged from 1 to 80 ft/d (Halford, 1998b, p. 26).

The water-table unit is recharged primarily by the infiltration of rainfall in the study area. Other sources of recharge include the land application of wastewater and reclaimed water, septic system effluent, irrigation in agricultural lands or residential areas, seepage from lakes and streams, and lateral groundwater inflow from adjacent areas. Recharge also can occur by means of upward leakage of water from the underlying limestone unit where the water levels are higher than in the water-table unit. Water is discharged from the water-table unit by pumping, evapotranspiration (ET), seepage into lakes, wetlands, or streams, lateral groundwater outflow to adjacent areas, and downward leakage of water from the water-table unit where the water levels are higher than in the limestone unit.

The elevation of the water table in the surficial aquifer system fluctuates in response to seasonal changes in precipitation and ET, and in response to pumping. Water levels are generally highest in September or October, at or near the end of the wet season and lowest in April or May, at or near the end of the dry season. Water-level fluctuations of up to 5 feet can occur between the wet and dry season. Along the St. Johns River, small fluctuations in water levels also can occur as the result of ocean tides. Water-level fluctuations of a few tenths

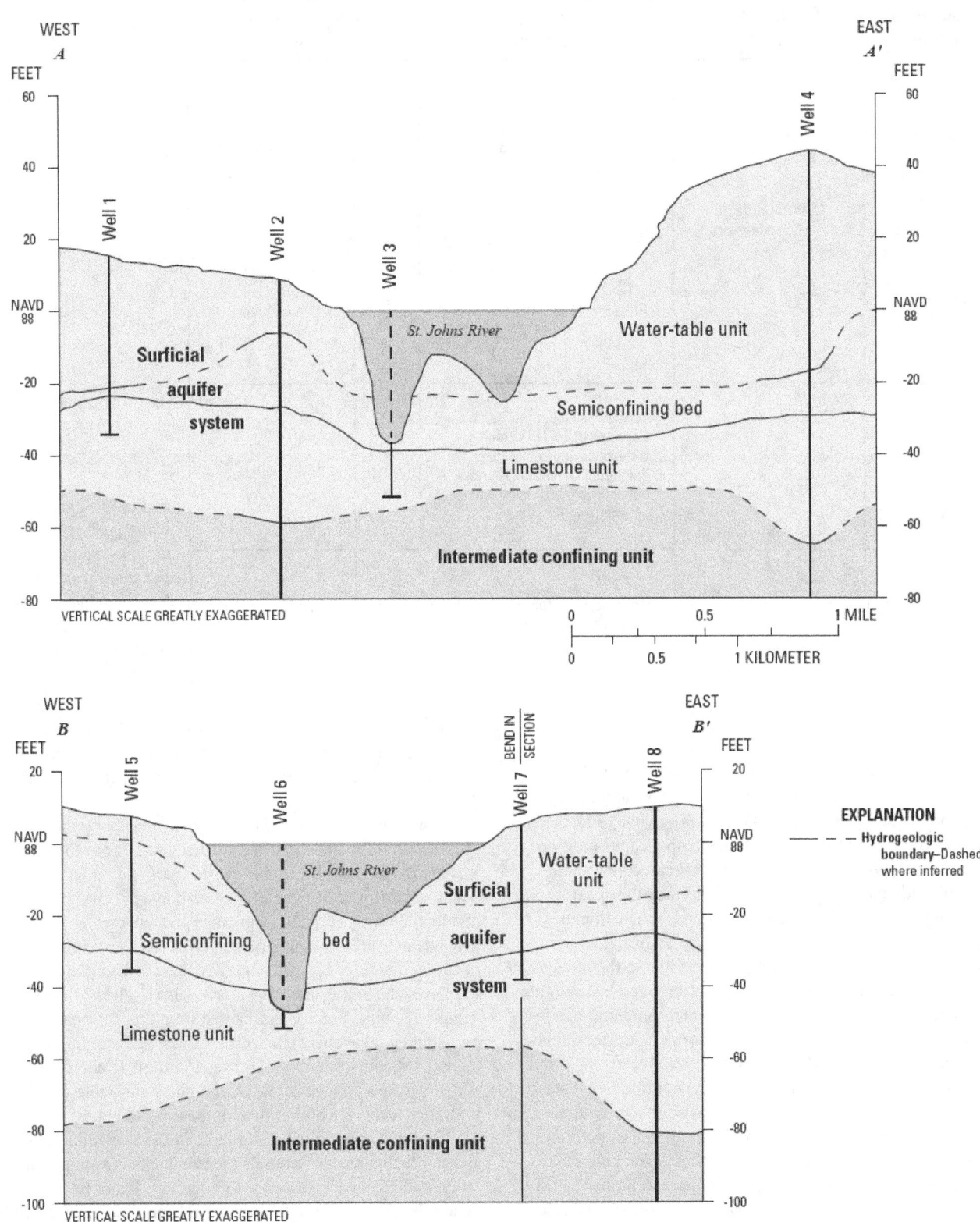

Figure 4. Generalized hydrogeologic sections *A–A'* and *B–B'*. Section lines shown in figure 1.

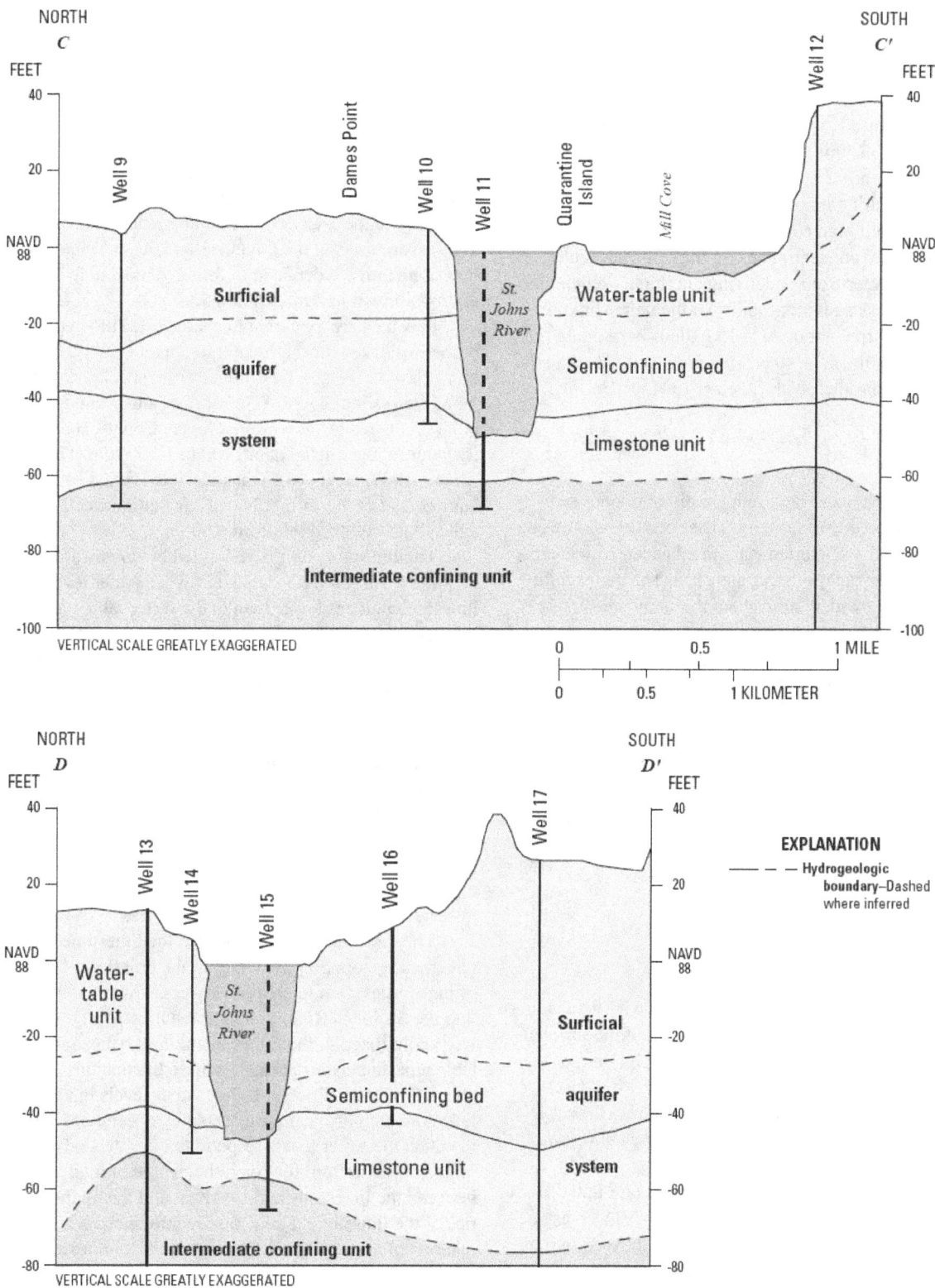

Figure 5. Generalized hydrogeologic sections *C–C'* and *D–D'*. Section lines shown in figure 1.

of a foot or less were observed in several monitoring wells drilled close to the river (Spechler and Stone, 1983). Halford (1998b, p. 29) also observed tidal effects in wells drilled in the water-table unit at Naval Station Mayport. Tidal fluctuations of water levels in 13 wells ranged from less than 0.02 to 0.9 foot and decreased rapidly away from the shoreline.

The water-table unit is not widely used as a source of water supply (except for lawn irrigation) because its permeability is low, resulting in low yields. In addition, water from this unit can contain high concentrations of dissolved iron or substantially higher concentrations of nutrients, pesticides, or bacteria than water from underlying aquifers. Well yields depend on thickness and permeability of the unit. Minimum well yields range from about 10 to 15 gallons per minute (gal/min) but can be as much as 40 gal/min from wells that tap the relatively permeable shell beds (Phelps, 1994, p. 28).

Semiconfining Bed

Underlying the water-table unit are discontinuous sedimentary beds of lower permeability. These beds consist of fine- to medium-grained, well-sorted sand, interbedded with layers of gray-green silty clay, clayey sand, and shell. The permeability of these beds varies widely throughout the study area. In areas where relatively little clay is present, the underlying limestone unit is unconfined, and the water-table and limestone units function as a single hydrologic unit (Causey and Phelps, 1978). In areas where the clay content of the semiconfining bed is substantial, the underlying limestone unit is under semiconfined or confined conditions. The thickness of these semiconfining beds, where present, ranges from about 5 to 40 feet. Along the navigation channel, much of the sediment that composes these beds has been removed by previous dredging projects. In some parts of the navigation channel, these sediments have been removed completely, exposing the limestone unit.

Limestone Unit

The limestone unit in the lower part of the surficial aquifer system is the principal water-yielding unit of the surficial aquifer system and is the focus of this study. The limestone unit consists mostly of limestone, interbedded with lenses of fine-to-medium sand, shell, and green calcareous silty clay. The limestone is poorly to moderately indurated, cavernous, sandy, and is dolomitic in places.

The elevation of the top of the limestone unit in the study area ranges from about 20 to 65 feet below NAVD 88, as indicated in figure 6. The top of the unit is shallowest in the extreme southwestern part of the study area near the Acosta Bridge and deepest in a small area in the northwestern part of the study area. The limestone is discontinuous in parts of the county and grades into medium- to coarse-grained sand and shell deposits along the coast. The unit ranges from about 5 to 40 feet in thickness.

The approximate elevation of the top of the limestone unit within the main navigation channel, based primarily on core borings provided by the USACE, generally ranges from about 39 to 54 feet below NAVD 88 and is slightly deeper in a few localized areas. These core borings and marine resistivity surveys also indicate that (1) the elevation of the top of the limestone in the navigation channel can vary as much as 6 or 7 feet within relatively short distances and (2) the unit is discontinuous throughout much of the navigation channel, especially in the area east of Fanning Island (Dexco, Inc., 2009). Variations in the thickness and depth to the top of the limestone unit are depicted in four generalized hydrogeologic sections shown in figures 4 and 5.

Few data are available to characterize the hydraulic properties of the limestone unit in Duval County. Estimated transmissivity values, determined from specific capacity tests, range from about 250 to 1,300 ft^2/d (Causey and Phelps, 1978, p. 20). At two sites in western Nassau County, transmissivity values ranged from about 200 to 1,000 ft^2/d (Dames and Moore, 1987, p. AI–16). Transmissivity values estimated from aquifer tests at the former Naval Air Station Cecil Field averaged about 800 ft^2/d (Halford, 1998b, p. 25).

The limestone unit is recharged by downward leakage of water from the water-table unit when water levels in the limestone unit are lower than in the water-table unit, and by groundwater inflow from adjacent areas. Recharge also can occur by upward leakage of water from the underlying Floridan aquifer system in areas where the water levels in the Floridan aquifer system are higher than in the limestone unit. In most of the study area, however, the actual rate of recharge to the limestone unit from the Floridan aquifer system may be low because the intermediate confining unit that separates the limestone unit from the Floridan aquifer system is thick and has low permeability. Water from the limestone unit is discharged primarily by pumping and by groundwater discharge to surface water, either through the water-table unit or directly into the St. Johns River and its tributaries.

The direction of lateral flow in the limestone unit is governed by topography. Water in the unit flows from areas of higher elevation to areas of lower elevation and discharges into the St. Johns River and its tributaries. Water levels in wells completed in the limestone unit, like those in the water-table unit, fluctuate seasonally; these fluctuations range from about 1 to 5 feet (Phelps, 1994). Water levels in wells from the limestone unit also can be affected by ocean tides. The degree to which the water levels respond to tides depends primarily on the distance from the river, the degree of interconnection between the limestone and the river, and the hydraulic properties of the limestone. Daily, tidally influenced water-level fluctuations of about 0.5 to 3 feet were observed in wells drilled at several locations close to the St. Johns River (Spechler and Stone, 1983).

Throughout much of the study area, the limestone unit is the principal water-producing unit of the surficial aquifer system. County-wide, domestic self-supply is the main use of water from this unit, which includes water used for drinking as

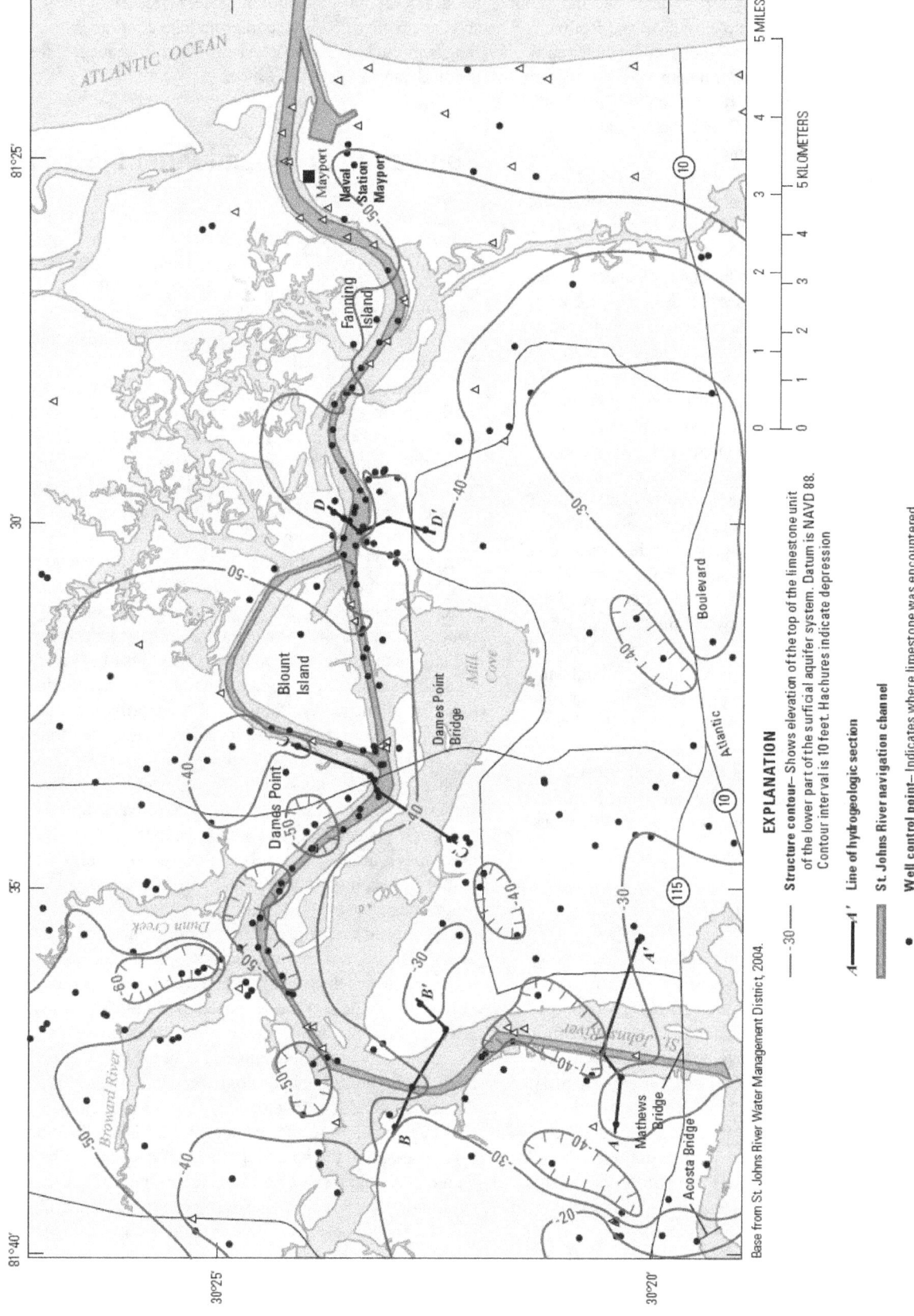

Base from St. Johns River Water Management District, 2004.

EXPLANATION

—30— **Structure contour**– Shows elevation of the top of the limestone unit of the lower part of the surficial aquifer system. Datum is NAVD 88. Contour interval is 10 feet. Hachures indicate depression

A——A' **Line of hydrogeologic section**

St. Johns River navigation channel

• **Well control point**– Indicates where limestone was encountered

△ **Well control point**– Indicates where limestone was not encountered

Figure 6. Generalized elevation of the top of the limestone unit of the lower part of the surficial aquifer system, modified from Spechler and Stone (1983).

well as lawn irrigation. In the study area, however, water from the limestone unit is used primarily for irrigation because public utilities supply drinking water to most of the area. Most wells completed in the limestone unit are 2 inches in diameter and generally yield about 30 to 100 gal/min, although yields of up to 200 gal/min have been reported (Causey and Phelps, 1978).

Throughout most of the study area, water quality in the limestone unit is sufficient for domestic, industrial, and commercial uses. Although the water from this unit can be hard, it generally meets State and Federal drinking-water standards (Fairchild, 1972; Causey and Phelps, 1978, Phelps, 1994). The quality of water withdrawn from the limestone unit along parts of the St. Johns River, near brackish-water marshes, and near the coast is less suitable for the uses just described. Several wells along the St. Johns River yielded water with salinities that ranged from 480 to 6,600 milligrams per liter (mg/L) chloride (0.9 to 11.9 ppt salinity) (Spechler and Stone, 1983).

A primary concern of this study is to determine whether the potential for saltwater intrusion is enhanced in areas where proposed dredging of the navigation channel will expose or excavate the limestone unit of the surficial aquifer system. Fractures and (or) joints (that is, secondary porosity) within the limestone are the primary flow pathways through the limestone unit; if an extensive network of highly transmissive fractures or joints becomes exposed to the river channel, the network could potentially transmit saltwater through the aquifer. The structure contours shown in figure 6 indicate that the limestone unit is between 40 and 50 feet below NAVD 88 along much of the river channel upstream of Fanning Island. This finding means that dredging operations could encounter the limestone unit over much of the segment that has been proposed for deepening. Because of the density-driven hydrodynamics of the river, saline water from the Atlantic Ocean travels upstream as a saltwater "wedge" along the bottom of the channel, precisely where the limestone is most likely to be exposed by the proposed dredging.

Intermediate Confining Unit

The intermediate confining unit includes all of the rock units that lie between the overlying surficial aquifer system and underlying Floridan aquifer system. Throughout the study area, the intermediate confining unit acts as a confining layer that restricts the vertical movement of water between the surficial aquifer system and Floridan aquifer system. The sediments have varying degrees of permeability, ranging from permeable limestone, dolostone, or sand to relatively impermeable layers of clay, clayey sand, or clayey carbonates. A generalized structural contour map of the elevation of the top of the intermediate confining unit is shown in figure 7. The top of the unit, which represents the base of the surficial aquifer system, ranges from about 30 feet below NAVD 88 in the extreme southwestern part of the study area to more than 100 feet below NAVD 88 in the northwestern part and is generally more than 70 feet below NAVD 88 north of the

river. The proposed deepening of the navigation channel will not affect the Floridan aquifer system because of the hydraulic separation provided by the intermediate confining unit, which ranges in thickness from more than 300 to about 500 feet in the study area (Spechler, 1996).

Simulation of Groundwater Flow

In this study, cross-sectional variable-density groundwater-flow and salinity transport models were developed using SEAWAT version 4 (Langevin and others, 2008) along four cross sections that intersect the St. Johns River navigation channel (fig. 8). The models were used to evaluate how changes to the river channel bathymetry may affect salinity in the surficial aquifer system.

The cross-sectional models are aligned along groundwater flow paths estimated from computed water-table-surface contours using the method described in Sepúlveda (2002) and in the discussion of head-dependent flux boundaries and no-flow boundaries herein. Cross-sectional models along specific transects were used rather than a single three-dimensional model of the entire study area because of time and resource constraints. The two-dimensional models were extended laterally beyond the banks of the St. Johns River to simulate saltwater movement from the river channel to the surficial aquifer system. These models provide a preliminary assessment of the effects of dredging on salinity variations in the surficial aquifer system. A three-dimensional model may provide a more comprehensive, although not necessarily more accurate, analysis of expected changes in salinity caused by dredging.

The models incorporate hydraulic properties, surficial aquifer system thicknesses, estimated water-table elevations, and groundwater withdrawal and recharge rates to determine a range of plausible aquifer responses to the proposed dredging. Models are usually calibrated to known conditions to reduce uncertainty. In this case, hydrologic conditions such as groundwater levels and salinity distribution were largely unknown, and as a consequence, these models are not calibrated. The uncertainty in hydrogeologic properties and hydrologic conditions therefore prevented an unambiguous simulation of the impacts of the proposed dredging. To estimate the effect of this uncertainty, the cross-sectional models were designed to incorporate a range of hydrogeologic conceptualizations and plausible hydrologic conditions. Thus, the two-dimensional models developed in this study do not necessarily simulate actual projected conditions, but rather, the models are used to examine the potential effects of deepening the navigation channel on saltwater intrusion in the surficial aquifer system under a range of plausible hypothetical conditions.

The study objectives necessitated the use of a tool to quantitatively simulate the physics of groundwater flow within an aquifer system that is subject to temporal and spatial variations in salinity. The tool must account for the effect of variable salinity concentrations in the aquifer system on

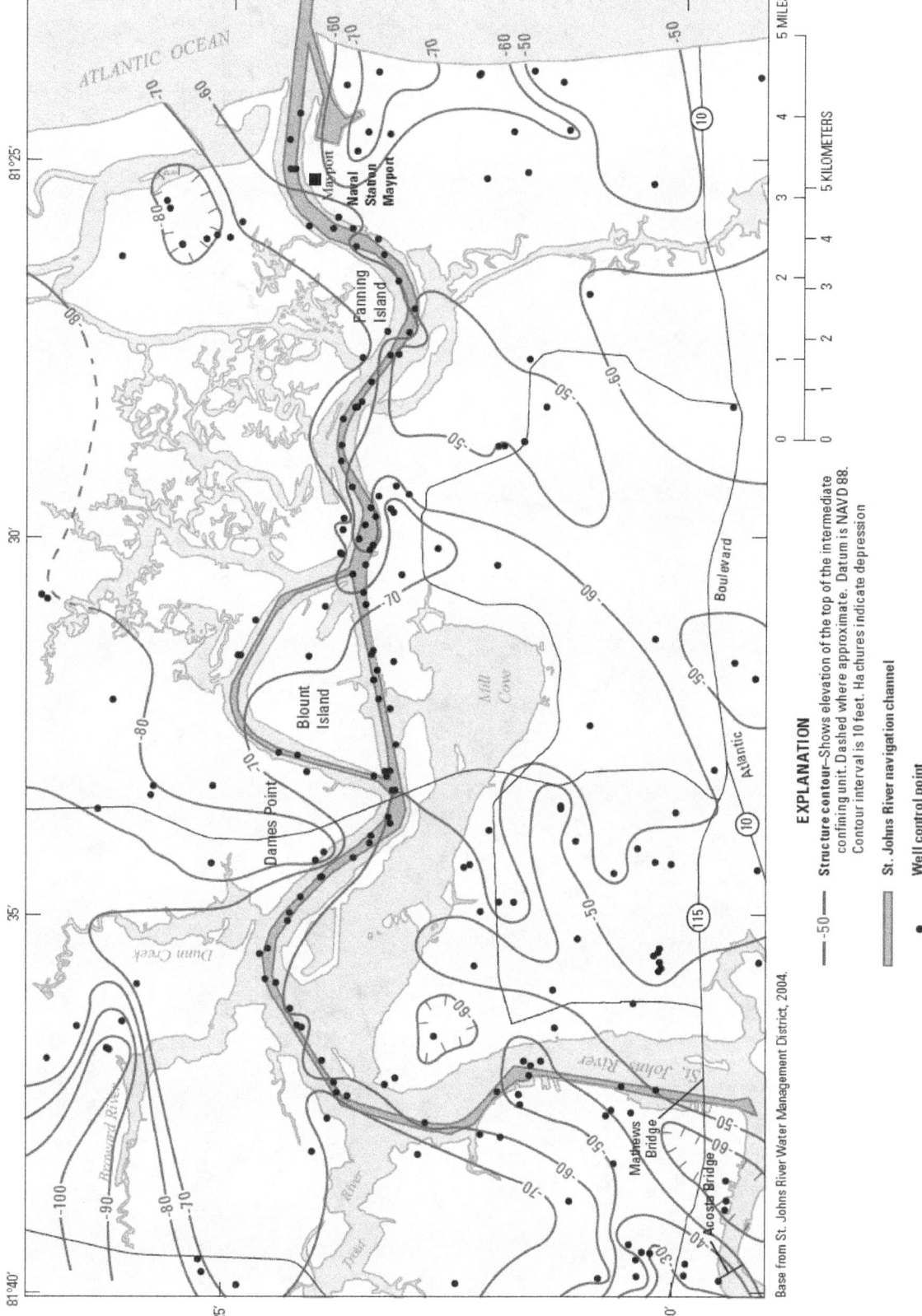

Base from St. Johns River Water Management District, 2004.

EXPLANATION

-50- Structure contour–Shows elevation of the top of the intermediate confining unit. Dashed where approximate. Datum is NAVD 88. Contour interval is 10 feet. Hachures indicate depression

St. Johns River navigation channel

• Well control point

Figure 7. Generalized elevation of the top of the intermediate confining unit.

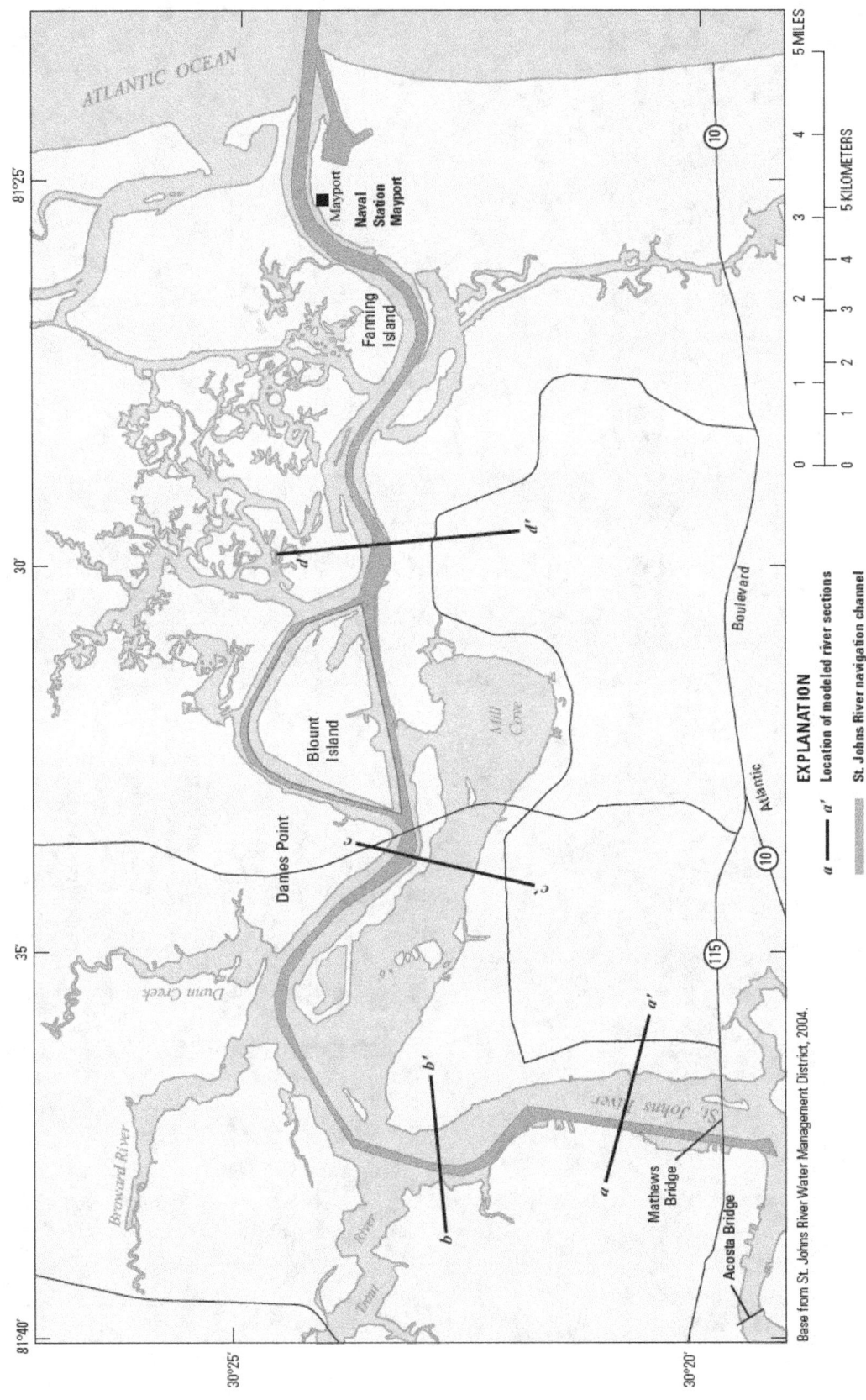

Base from St. Johns River Water Management District, 2004.

Figure 8. Location of modeled river sections.

water-density variations and, therefore, on the flow system. This tool solves the variable-density groundwater-flow and solute-transport equations incorporating hydraulic parameters and initial/boundary conditions representative of the study area. The SEAWAT computer program for simulation of variable-density groundwater flow (Guo and Langevin, 2002; Langevin and others, 2003; Langevin and others, 2008) was designed to implement this solution by combining a modified version of the groundwater-flow model MODFLOW (McDonald and Harbaugh, 1988; Harbaugh and McDonald, 1996) and the solute-transport model MT3DMS (Zheng and Wang, 1998). Additional information about the mathematical formulation, benchmark testing, and application of MT3DMS, MODFLOW, and SEAWAT are provided in Zheng and Wang (1998), Harbaugh and others (2000), Guo and Langevin (2002), Langevin and others (2003), and Langevin and others, (2008). SEAWAT version 4.00.05 was run in 64-bit memory space for all model simulations described herein.

Model Design

The two-dimensional variable-density groundwater-flow models were developed at four cross sections along the St. Johns river from near River Mile 20 to near River Mile 8 (fig. 8) in areas in which the surficial aquifer system was thought to be most vulnerable to saltwater intrusion (fig. 8). Models were oriented along estimated groundwater flow lines and extended inland as far as possible. Saltwater movement within the groundwater system was evaluated for dredged and undredged channel configurations. Model input data include geologic surface elevations, net groundwater recharge, groundwater pumpage, river stage, and salinity concentrations.

The impact of channel dredging on variations in salinity in the groundwater system was evaluated for the 121-day period from December 1, 1998, through March 31, 1999, to coincide with the available river stage and salinity results from the EFDC (Hamrick, 1992) hydrodynamic model simulation for the St. Johns River provided by the USACE (Steve Bratos, U.S. Army Corps of Engineers, written commun., 2012). Extended 363-day model simulations were run to determine whether peak salinity differences were reached during the 121-day simulations. Input datasets for 363-day simulations were created by inserting one mirror-image of the 121-day data between two unmanipulated copies of the data. This approach was necessary to create a smooth extended time series because the first value of the reversed dataset aligns with the last value of the first dataset and the last value of the reversed dataset aligns with the first value of the last dataset. Without the reversal of the middle dataset, discontinuities in the time series would prevent convergence of extended model simulations. In both sets of simulations, transient model simulations used daily stress periods as well as identical model grids, aquifer properties, and boundary conditions. Hourly stress periods, which capture tidal fluctuations, were evaluated and model results were not appreciably different than those of models using daily stress periods.

Spatial Discretization

Spatial discretization of the finite-difference model grids was guided by the need to adequately define hydraulic gradients and hydrogeologic features, subject to computational constraints. In each two-dimensional cross-sectional model, model cells are 16.4 feet long (5 meters, lateral dimension) by approximately 3.3 feet wide (1 meter, lateral dimension), and 3.3 feet thick (1 meter, vertical dimension). The number of columns varies among the models depending on the width of the river channel and configuration of the potentiometric surface. A total of 708, 647, 769, and 1,030 columns were used in cross-sectional models a–a', b–b', c–c', and d–d' (fig. 8), respectively.

Variable-density groundwater-flow systems are affected by vertical density gradients and require the use of finer vertical model discretization, relative to constant-density groundwater-flow models, to simulate these systems accurately. A total of 30, 30, 40, and 50 layers were used in cross-sectional models a–a', b–b', c–c', and d–d', respectively. The top and bottom of each cross section correspond to the highest land surface elevation and the lowest elevation of the bottom of the limestone unit, respectively.

Aquifer Properties

Uniform parameter values were used to represent the hydraulic properties of each discrete layer or feature of the surficial aquifer system. Horizontal hydraulic conductivity values of the water-table and limestone units are based on those reported by Causey and Phelps (1978), Davis and others (1996), and Halford (1998a, b). Solution-enhanced features have been observed in parts of the limestone unit and could increase the potential for saltwater intrusion, especially if dredging activities were to expose them in the river channel. These features have been represented in some of the model simulation pairs (described subsequently) as high hydraulic conductivity layers with horizontal and vertical hydraulic conductivities of 450 ft/d, based on data from Causey and Phelps (1978). Semiconfining beds separating the water-table unit from the limestone unit were represented in some model simulation pairs as low hydraulic conductivity layers. No field-derived hydraulic conductivity values were available for these beds; horizontal hydraulic conductivity was assumed to be half that of the water-table unit, and vertical hydraulic conductivity was assumed to be one-tenth of the assumed horizontal conductivity value.

Specific storage (S_s) and specific yield (S_y) parameters are used to describe the amount of water that an aquifer releases from storage when head in the aquifer declines. Specific storage refers to the confined case and is expressed in terms of volume per unit decline in head. Specific storage values used in the models were $5.0×10^{-5}$, $1.0×10^{-4}$, and $1.0×10^{-3}$ ft^{-1} for the limestone unit, water-table unit, and semiconfining beds, respectively. Specific yield refers to the unconfined case and is expressed in terms of volume per unit surface area per unit decline in head. Specific yield values used in the models were 6.0×10^{-2}, 3.2×10^{-1}, and 1.4×10^{-1} for the semiconfining beds, water-table unit, and limestone unit, respectively. Both S_s and S_y values were specified to allow model layers to convert from confined to unconfined conditions if needed. Values were based on those found in Anderson and Woessner (2002, tables 3.4 and 3.5).

Effective porosity and hydrodynamic dispersion parameters are necessary to simulate solute transport through a porous medium (Langevin, 2001). Effective porosity represents the porosity available for fluid flow and is conceptually similar to specific yield. For this study, effective porosity was assumed to be equivalent to specific yield. Hydrodynamic dispersion represents the combined effects of molecular and mechanical dispersion. Molecular dispersion was not simulated in this study because it was assumed to be minimal compared to mechanical dispersion. Mechanical dispersion is a result of aquifer heterogeneities that result in complex velocity fields that are typically not represented explicitly in groundwater-flow and transport models (Konikow, 2011). Longitudinal and transverse (horizontal and vertical) dispersivity were specified as 16 and 1.6 feet, respectively, and represent the averages used in two cross-sectional, variable-density models developed by Langevin (2001) for the limestone aquifers underlying Biscayne Bay in southern Florida.

Boundary Conditions

The solution of the groundwater-flow and transport equations inherent in the model requires the specification of boundary conditions. Three common types of boundary conditions are specified to solve the flow and transport equations of numerical models: specified head (Dirichlet boundary); specified flux (Neumann boundary), and head-dependent flux (Cauchy or mixed-boundary) (Reilly, 2001). In groundwater-flow and transport models, boundary conditions represent heads and solute concentrations in physical features, such as lakes, rivers, and wells, and fluxes from processes and features, such as ET, recharge, and groundwater divides. The groundwater divides are no-flow boundaries, a special case of specified-flux boundaries. For transient models, initial water levels and solute concentrations in the surficial aquifer system were specified using results from steady-state simulations.

Groundwater Withdrawals

Groundwater withdrawals are a type of specified-flux boundary and were applied to models in this study using the well (WEL) package. Withdrawal of water was treated in the models as a diffuse sink, similar to negative recharge, in which the relatively small amounts of pumpage are computed in units of length per time and distributed over a grid throughout the study area. Water was removed from models by including a well in each model cell for which the withdrawal rate was greater than zero; volumes were computed by multiplying the withdrawal rate times the model cell area. Diffuse

groundwater withdrawals may be simulated using the recharge (RCH) package, but the WEL package was used because it allows specified flux to be applied to multiple layers and both flow units in the surficial aquifer are pumped in the study area.

Withdrawals from the surficial aquifer system in the study area were estimated using land-use and census data as well as 2010 maps of public-water-supply service areas (fig. 9) and were assumed to be constant throughout the transient simulation period. Land-use data were used to identify those areas in which groundwater pumpage from the surficial aquifer system for lawn irrigation was likely to be greatest. These data were intersected with 2010 Census data (U.S. Census Bureau, 2012) to determine population density and estimate groundwater pumpage from the surficial aquifer system in the study area. Estimates of groundwater pumpage from the surficial aquifer system were based on two assumptions: (1) groundwater pumpage is only for lawn irrigation in public-water-supply service areas, and (2) no groundwater pumpage occurs in high-density residential areas.

The majority of the study area, except for unpopulated areas, is within public-water-supply service areas. As a result, all water used in the study area was assumed to be for domestic lawn irrigation, which accounts for 2.75 Mgal/d of the 8.8 Mgal/d withdrawn from the surficial aquifer system (Richard L. Marella, U.S. Geological Survey, written commun., 2011). It is estimated that of the 115,500 domestic wells in Duval County public-water-supply service areas, 25 percent (28,875 wells) withdraw water from the surficial aquifer system (Richard L. Marella, U.S. Geological Survey, written commun., 2011). Dividing this by the total withdrawals for domestic irrigation (2.75 Mgal/d) yields an average of 95.2 gallons per day (gal/d) per surficial aquifer system well in Duval County public-water-supply service areas.

Housing unit data from the 2010 Census (U.S. Census Bureau, 2012) were used to distribute the withdrawal of groundwater by assuming that each housing unit has one well. These wells were then aggregated using a 1,576×1,291-foot (480×394-meter) regular grid covering the study area. The total number of wells in each cell includes wells in both the Floridan and surficial aquifer systems because some wells in public-water-supply service areas do not withdraw water from the surficial aquifer system. To account for this, a well factor was introduced to reduce the number of wells in each cell. In areas where all wells were assumed to withdraw from the Floridan aquifer system, the well factor was assigned a value of zero. A well factor of 0.25 was assigned in the remaining areas where 25 percent of the wells were assumed to withdraw from the surficial aquifer system. Total groundwater pumpage for domestic self-supplied irrigation purposes was estimated to be 1.0 Mgal/d based on 10,900 surficial aquifer system wells in the study area pumping at a rate of 95.2 gal/d.

A well dataset from the City of Jacksonville Environmental Quality Division provided the basis for determining the overall distribution of groundwater pumping between the water-table and limestone units of the surficial aquifer system. Although this dataset is incomplete, it was assumed to contain

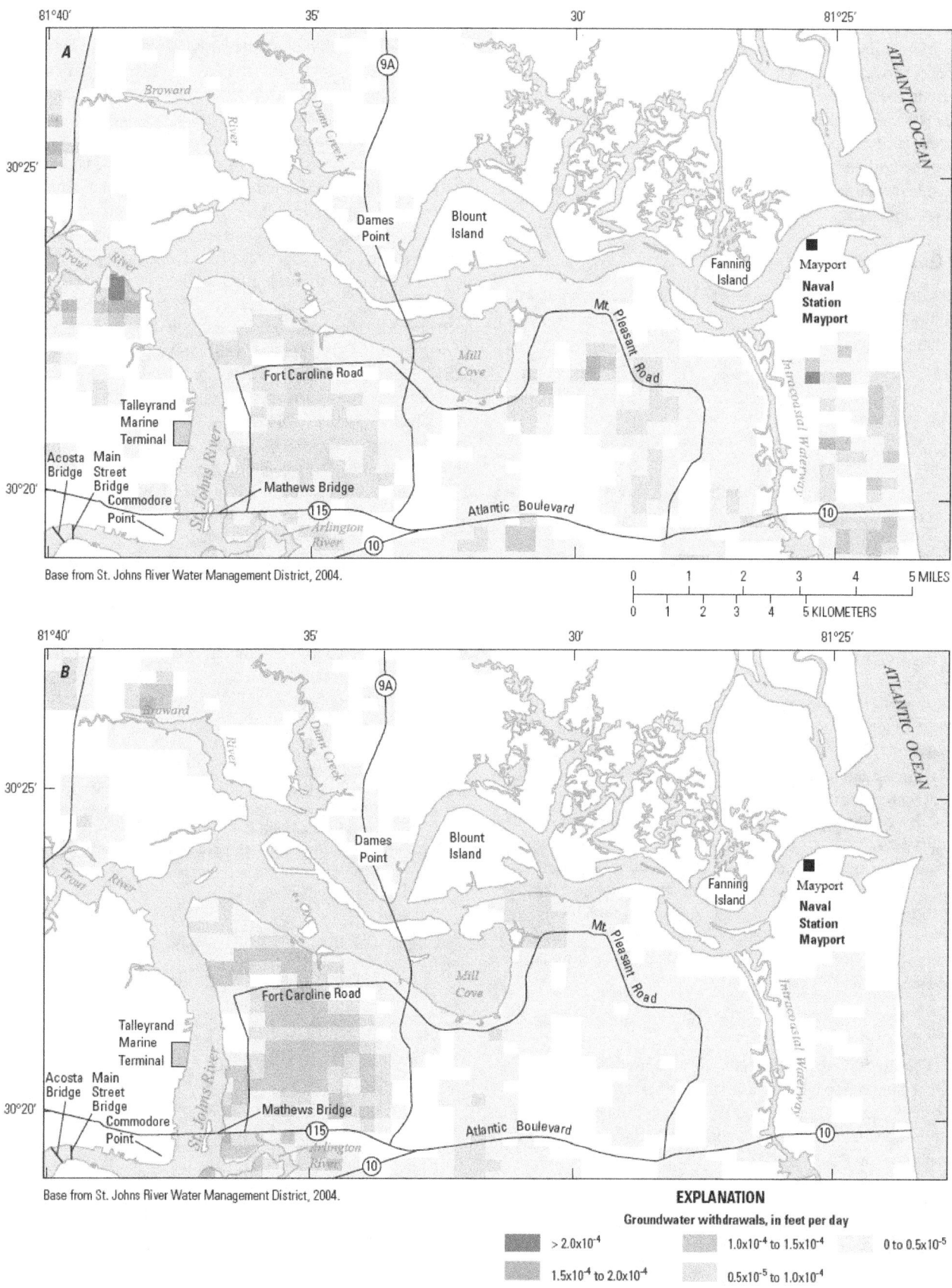

Figure 9. Estimated groundwater withdrawal rates from *(A)* the water-table and *(B)* limestone units of the surficial aquifer system, 2011.

a representative sample of wells in the study area. The proportion of water-table to limestone unit wells ranges from 23 to 38 percent over much of the western half of the study area, except in the downtown area where all wells are water-table wells (100 percent). In the southeastern portion of the study area, the proportion ranges from 80 to 91 percent where the limestone unit becomes discontinuous. The average proportion of water-table to limestone unit wells over the entire study area is 52 percent.

Net Groundwater Recharge

Net groundwater recharge, a specified-flux boundary condition, was implemented using the RCH package and was calculated using the following equation:

$$R = P(1-C) + Q_{septic} - ET \qquad (1)$$

where

R	is net groundwater recharge $[LT^{-1}]$,
P	is precipitation $[LT^{-1}]$,
C	is the runoff coefficient [unitless],
Q_{septic}	is discharge from septic systems $[LT^{-1}]$, and
ET	is evapotranspiration $[LT^{-1}]$.

Net groundwater recharge was calculated for a grid of 3,600 cells, each about 1,427×1,427 feet (435×435 meters) in area; values were extracted from this grid and applied to the steady-state models. Average net groundwater recharge over the entire study area was calculated to be 11.5 inches per year (in/yr) and was applied in the steady-state models. Daily net groundwater recharge was computed from December 1, 1998, through April 1, 1999, and was applied in the transient models; total net groundwater recharge during this period was −0.7 inch. Negative total net groundwater recharge indicates a loss of water from the system, caused in this case by abnormally dry conditions in which the rate of ET exceeded that of precipitation. Previous modeling studies have used hydrograph separation techniques and groundwater model calibration to estimate net groundwater recharge; these results range from 4 to 14 in/yr (Halford, 1998a, b). Locations of model cell nodes were used to extract values from the net groundwater recharge grid. Derivations of the values input into equation 1 are described next.

Precipitation and Evapotranspiration

The PRISM (Parameter-elevation Regressions on Independent Slopes Model) Climate Group at Oregon State University maintains a high-resolution precipitation dataset for the conterminous United States, available as monthly totals from 1895 to present (Daly and Gibson, 2002; Daly and others, 2011). Average annual rainfall in the study area for the 16-year period from January 1983 to December 1998 was 53.7 inches, and monthly averages ranged from 2.3 inches in May to 7.8 inches in September (fig. 10). Average annual rainfall was used in the computation of net groundwater recharge for the steady-state models. Precipitation data from the Next Generation Weather Radar system (NEXRAD) digital precipitation

array dataset maintained by the National Climatic Data Center (NCDC) were used to compute daily rainfall totals used in the transient models. Total NEXRAD rainfall in the study area for the period from December 1, 1998, through March 31, 1999, was substantially below average, totaling 3.42 inches instead of the periodic average of 12.44 inches computed from rainfall data collected at National Weather Service (NWS) cooperative observer station 84358 (Jacksonville International Airport) from 1938 through 2013 (fig. 1). Additionally, recorded rainfall at Jacksonville International Airport from December 1998 through March 1999 was exceeded in nearly 89 percent of like periods from 1938 through 2013.

Reference evapotranspiration (RET) is the ET that would occur from a standard reference crop, defined as an actively-growing crop of green grass or alfalfa with no soil water deficit. Daily statewide RET data for Florida are available at a 2-kilometer (1.24-mile) resolution (Jacobs and others, 2008; Mecikalski and others, 2011; U.S. Geological Survey, 2012) and were used in this study. Varying meteorological conditions, rooting zone moisture availability, and plant phenology affect the amount of water transpired by a crop, with maximum ET during the spring and summer growing season and minimum ET during the winter (dormant) season (Jia and others, 2009). Crop coefficients (K_c) are typically used to scale RET data for nonreference land covers or soil moisture conditions; for this study, K_c values for Bahia grass, a warm weather grass common in Florida, were used to calculate ET in the study area (Jia and others, 2009). Crop coefficient values ranged from 0.35 in January and February to 0.9 in May. Calculated ET was multiplied by 1 minus a runoff coefficient to account for reduced ET

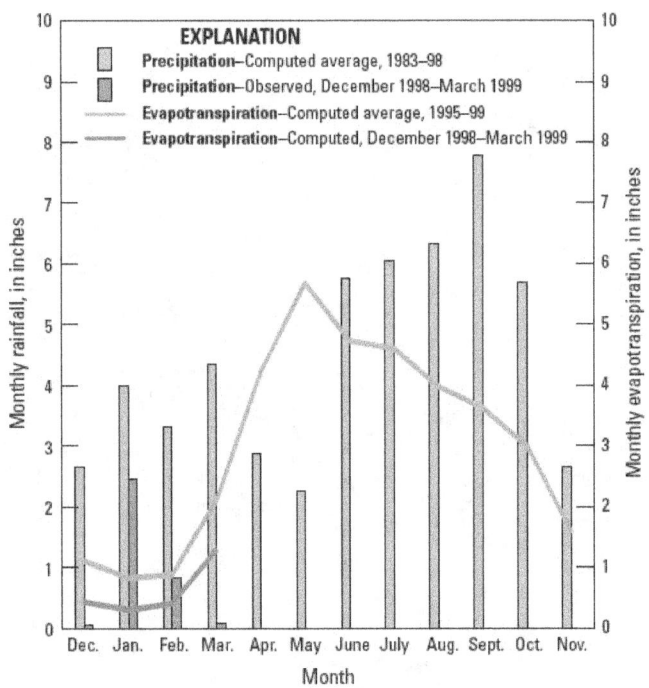

Figure 10.   Monthly precipitation and evapotranspiration in the study area.

in areas where vegetation was sparse or nonexistent, such as roadways, industrial areas, and open bodies of water. December–March monthly average ET during 1995–99 ranged from 0.9 inch in January to 2.4 inches in March. Monthly ET from December 1998 through March 1999 ranged from 0.3 inch in January to 1.3 inches in March. ET was not explicitly simulated in the models, but calculated ET rates were used to develop net groundwater recharge data. Figure 10 shows the 1983–99 monthly average computed ET and December 1998 to March 1999 computed monthly ET values in the study area.

Surface-Water Runoff

A runoff coefficient was used to represent the proportion of precipitation that remains above land surface and accumulates in surface-water bodies rather than infiltrating the soil and recharging the water table. The proportion of runoff generated is related to land use and land-surface slope; because the study area is generally flat, runoff coefficient values were assigned based on land use alone and calculated using data from Fetter (2001, table 2.3). Runoff coefficient values ranged from 0.1 for the open lands and forest, recreation, and agriculture land-use categories to 0.9 for nonpervious land uses in the commercial category, such as roads, parking lots, and so forth. Lake, marshes, and other water bodies that are internally drained were assigned a runoff coefficient of 0. Rivers, coastal marshes, and other water bodies that drain to the ocean were assigned a runoff coefficient of 1.

Septic Systems

Discharge to the surficial aquifer system from septic systems was calculated using a dataset compiled by the Florida Department of Health (FDOH) (Hall and Clancy, 2009). Septic systems were identified using an algorithm that uses tax records from the Florida Department of Revenue (DOR) and a regression based on available septic system data and utility records to compute the probability that a given improved parcel has an active septic system. Parcels having a probability greater than or equal to 0.5 were considered to have an actively discharging septic system, the location of which was assumed to be the parcel centroid. These septic system locations were intersected with data from the 2010 Census (U.S. Census Bureau, 2012) to estimate the rate of domestic wastewater discharge to each active septic system based on population density. In 2009 there were 27,131 septic systems in the study area, and the mean population per household in areas with septic systems was 2.8. Using an average discharge of 55 gal/d per person (Marella, 2004), the total discharge to septic systems in the study area was 3.9 Mgal/d. All water supplied for residential use in homes with a septic system was assumed to ultimately recharge the water table.

Head-Dependent Flux Boundaries and No-Flow Boundaries

Head-dependent flux (Cauchy) boundaries allow flow into and out of the model from an external source. The flow is proportional to the difference between the specified boundary head and the calculated head. Several head-dependent flux boundary condition packages are available in SEAWAT (Langevin and others, 2008).

The river (RIV) package was used to simulate the exchange of water between the St. Johns River channel and the surficial aquifer system. For the steady-state models, stage in the St. Johns River was set to –0.33 foot NAVD 88, equivalent to average river stage computed from historical data collected at USGS streamflow gaging station 02246500 (St. Johns River at Jacksonville, FL) (U.S. Geological Survey, 2013c) (fig. 1, table 1). For the transient models, St. Johns River stage values were specified using simulated hydrodynamic model results provided by the USACE (Steve Bratos, U.S. Army Corps of Engineers, written commun., 2012). RIV package conductance values were calculated using the equation:

$$C = \frac{K_v A}{L} \tag{2}$$

where
- C is the conductance [LT^{-1}],
- K_v is the vertical hydraulic conductivity [LT^{-1}],
- A is the area of the cell face through which flow occurs [L^2], and
- L is the distance between nodes [LT^{-1}].

Boundary conditions along the end of each cross section were specified using the general head boundary (GHB) package. GHBs were added to simulate inflow to the model domain from upgradient areas not explicitly simulated in the models. The GHB package conductance values were calculated by substituting horizontal hydraulic conductivity (K_h) (water-table unit: 5 ft/d; limestone: 40 ft/d; semiconfining beds: 3 ft/d; and preferential flow paths: 656 ft/d) for K_v in equation 2. GHB heads did not vary over time and were based on estimated water-table elevations calculated using the multiple-linear-regression method developed for peninsular Florida by Sepúlveda (2002). This method introduces the concept of the minimum water table, which is the elevation of the water table at aquifer drain features, such as streams and lakes. The equation used to estimate the mean annual elevation of the water table is:

$$WT_i = \beta_1 MINWT_i + \beta_2 (LSA_i - MINWT_i) \tag{3}$$

where
- WT_i is the calculated water-table elevation at cell I,
- $MINWT_i$ is the minimum water-table elevation interpolated at cell I,
- LSA_i is the land surface elevation interpolated at cell I, and
- β_1 and β_2 are the dimensionless regression coefficients of the multiple linear regression.

Land surface elevations were taken from 3-meter (9.84-foot) resolution light detection and ranging (lidar) data from the USGS National Elevation Dataset (NED) (Gesch and others, 2002; Gesch, 2007). Regression coefficients β_1 and β_2 are assigned from table 2 of Sepúlveda (2002) and are based on physiographic region groups.

Cells above land surface or above the channel bottom were specified as no-flow cells. The bottom of the model is also a no-flow boundary because exchanges between the surficial aquifer system and underlying intermediate confining unit are assumed to be minimal.

Specified Solute Concentrations

The source and sink mixing (SSM) package for SEAWAT is used to specify solute concentrations for specified groundwater boundary conditions. Net groundwater recharge and GHBs at the edge of the model domains were assumed to have a concentration of 0 ppt, except for GHBs located in the salt marsh at the northern boundary of model cross section d–d' where average daily salinity and river-stage values from the USACE EFDC model (hereafter, EFDC model) were specified. Solute concentrations for RIV package cells varied along each cross section and were specified using results from the EFDC model simulations.

Initial Water Levels and Solute Concentrations

Initial water levels and solute concentrations are considered initial conditions and are used by the backwards difference approximation method of SEAWAT to calculate head and solute concentrations for the first time step of transient simulations. As stated earlier herein, initial conditions for transient models were specified using results from steady-state simulations. Initial water levels for steady-state model simulations in this study were specified using computed values to decrease the time needed to reach dynamic equilibrium. Limited field data were available from the USGS National Water Information System (NWIS) database for water levels in the surficial aquifer system in Duval County; however, the spatial and temporal distribution of the data was not sufficient to accurately map the water-table elevation at a scale appropriate for this study. Initial water levels for steady-state simulations were calculated using equation 3. The solute concentration data available for this study were not sufficient to accurately describe the complex distribution of salinity along the bottom of the river channel. Simulation results from the EFDC model were used to specify initial salinity concentrations for steady-state simulations. The salinity of the entire surficial aquifer system was set to equal the average salinity of the bottom layer of the EFDC model at each model cross section; values ranged from 18.3 ppt for model a–a' to 30.5 ppt for model d–d'. Steady-state model simulations required 15 to 65 years to allow freshwater to flush through the system before salinity values at each observation point ceased to change with time, indicating dynamic equilibrium was reached.

Combined steady-state and transient simulation pairs were executed for each simulation run; details about the simulation pairs used are described later herein. Figure 11 shows a

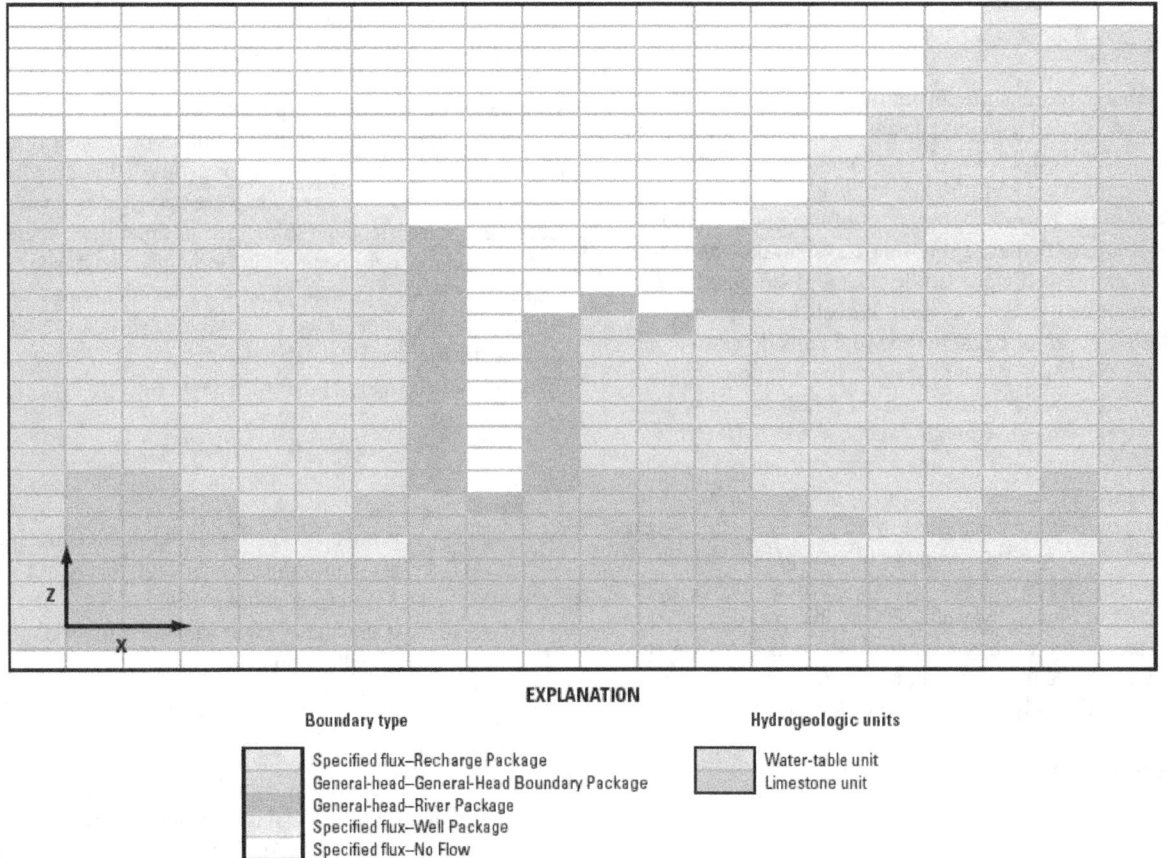

EXPLANATION

Boundary type
Specified flux–Recharge Package
General-head–General-Head Boundary Package
General-head–River Package
Specified flux–Well Package
Specified flux–No Flow

Hydrogeologic units
Water-table unit
Limestone unit

Figure 11.   Finite difference model grid and boundary conditions.

generalized finite-difference model grid that incorporates the boundary conditions applied to models used for this study.

Model Simulations

Simulation pairs (table 2, fig. 12) were run for each cross-sectional model to determine how several conceptualized hydrogeologic structures affect the computed salinity distribution in the surficial aquifer system. Each pair consisted of one simulation performed with the existing, undredged channel bathymetry and the other with the dredged channel bathymetry. Simulation pairs used one of four hydrogeologic realizations, which ranged from simple to complex to convey the range of uncertainty in hydrogeologic characterization of the system. The simplest realization includes only the sandy water-table unit and a competent limestone unit. Alternative scenarios include the complex hydrogeologic structures of a contiguous, 20-foot-thick semiconfining bed and (or) a high hydraulic conductivity (K) preferential flow path in the limestone (fig. 12). Both of these structures were tested individually and then together to simulate a range of possible hydrogeologic models.

Table 2. Number, name, and description of model simulation pairs.

[NAVD 88, North American Vertical Datum of 1988]

Simulation pair number (fig. 12)	Simulation pair name	Description
1	Simple system	Surficial aquifer system consists of upper water-table unit and limestone unit only.
2	Semiconfined system	Surficial aquifer system consists of upper water-table unit and limestone unit separated by a 20-foot-thick semiconfining bed.
3	Preferential flow system	Surficial aquifer system consists of upper water-table unit and limestone unit with a high-hydraulic-conductivity preferential flow layer.
4	Complex system	Surficial aquifer system consists of upper water-table unit and limestone unit with a high-hydraulic-conductivity preferential flow layer. Units are separated by a 20-foot-thick semiconfining bed.

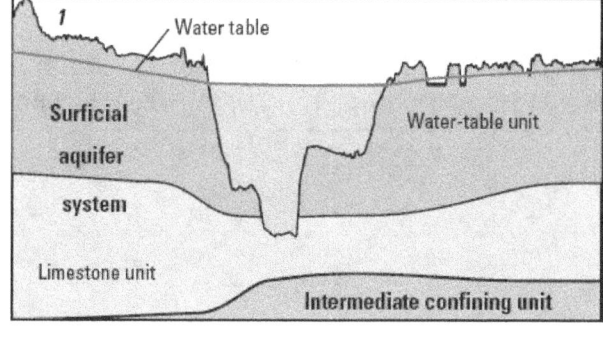

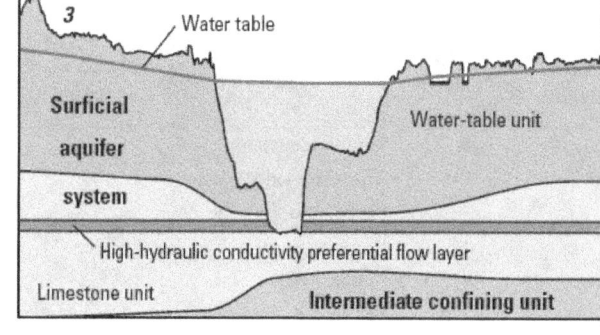

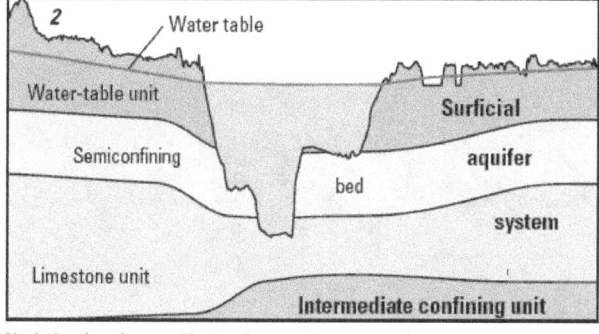

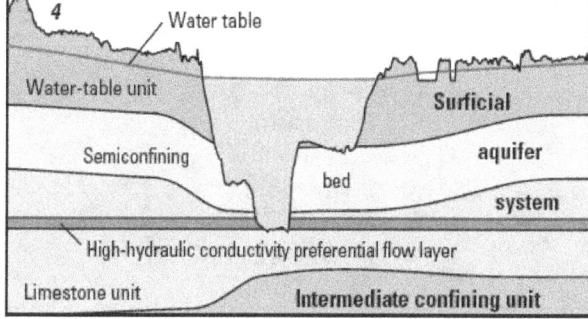

Vertical scale and water table elevations greatly exaggerated

Figure 12. Model simulation pairs at model cross section b–b'. Simulation pair numbers are listed in the upper left-hand corner of each plot; names and descriptions are given in table 2.

Model Results

Model results include groundwater budgets for each cross section. These water budgets indicated a net discharge of water from the aquifer to the river for all tested scenarios at all cross sections. Figures 13 and 14 show the average, undredged 0.45-ppt (250 mg/L chloride) and 18.1-ppt (10,000 mg/L chloride) salinity contours for each of the four cross-sectional models. The 0.45-ppt contour is shown because it represents the national secondary drinking-water standard of 250 mg/L for chloride (U.S. Code of Federal Regulations, 2002). Simulation results for each cross-sectional model showed a mass of relatively more saline water beneath the channel surrounded by relatively less saline water. The elevated salinities in these areas, indicated by the simulation results, are the result of saltwater that remained after completion of the steady-state simulations establishing the initial conditions, and possibly downward leakage of denser, more saline water from the river into the aquifer.

Results were recorded at eight observation points for each simulation. Observation points were specified to be about 75 and 730 feet from the channel on both sides of the river and include a shallow observation point (SH) in the water-table portion of the aquifer and a deep observation point (DP) in the limestone portion of the aquifer. Depths of observation points varied among cross sections depending on the thickness of the

units (figs. 13 and 14). Changes is salinity were calculated for each simulation pair by subtracting undredged values from dredged values for each observation point at the end of each time step. Results of most simulation pairs indicated limited salinity increases for the conditions represented by these models (table 3). Dredging had no substantial effect on groundwater salinity along model cross sections *a–a'*, *b–b'*, or *c-c'* for any of the simulation pairs (all salinity increases less than 0.05 ppt). Results for these cross sections are not discussed further herein.

Results of simulations for modeled conditions indicate that dredging will have little or no effect on salinity at SHs in the water-table unit along model cross section *d–d'* and increases in computed salinity were generally modest at deep observation points in the limestone unit. The largest increases in computed salinity were for the complex system simulation pair, which indicated maximum increases of 3.7 and 4.0 ppt at observation points DP–1 and DP–2, respectively. Maximum increases for the majority of other simulation pairs were less than 0.5 ppt.

Time-series graphs of results at deep observation points for all simulation pairs at cross section *d–d'* (fig. 15) depict salinity increases over the 121-day transient period. In some cases, trends indicate that salinity increases at DP–2 continued to rise beyond the end of the simulation period.

Table 3. Maximum computed salinity increases at observation points caused by proposed dredging.

[ppt, parts per thousand; SH, shallow observation well; DP, deep observation well; --, value less than 0.05 parts per thousand]

Cross section	Simulation pair number (fig. 12)	Simulation pair name	Maximum computed salinity differences, in ppt							
			SH-1	SH-2	SH-3	SH-4	DP-1	DP-2	DP-3	DP-4
a–a'	1	Simple system	--	--	--	--	--	--	--	--
	2	Semiconfined system	--	--	--	--	--	--	--	--
	3	Preferential flow system	--	--	--	--	--	--	--	--
	4	Complex system	--	--	--	--	--	--	--	--
b–b'	1	Simple system	--	--	--	--	--	--	--	--
	2	Semiconfined system	--	--	--	--	--	--	--	--
	3	Preferential flow system	--	--	--	--	--	--	--	--
	4	Complex system	--	--	--	--	--	--	--	--
c–c'	1	Simple system	--	--	--	--	--	--	--	--
	2	Semiconfined system	--	--	--	--	--	--	--	--
	3	Preferential flow system	--	--	--	--	--	--	--	--
	4	Complex system	--	--	--	--	--	--	--	--
d–d'	1	Simple system	--	--	--	--	0.2	1.3	--	--
	2	Semiconfined system	--	--	--	--	--	0.1	--	--
	3	Preferential flow system	--	--	--	--	0.2	0.4	--	--
	4	Complex system	--	--	--	--	3.7	4.0	--	--

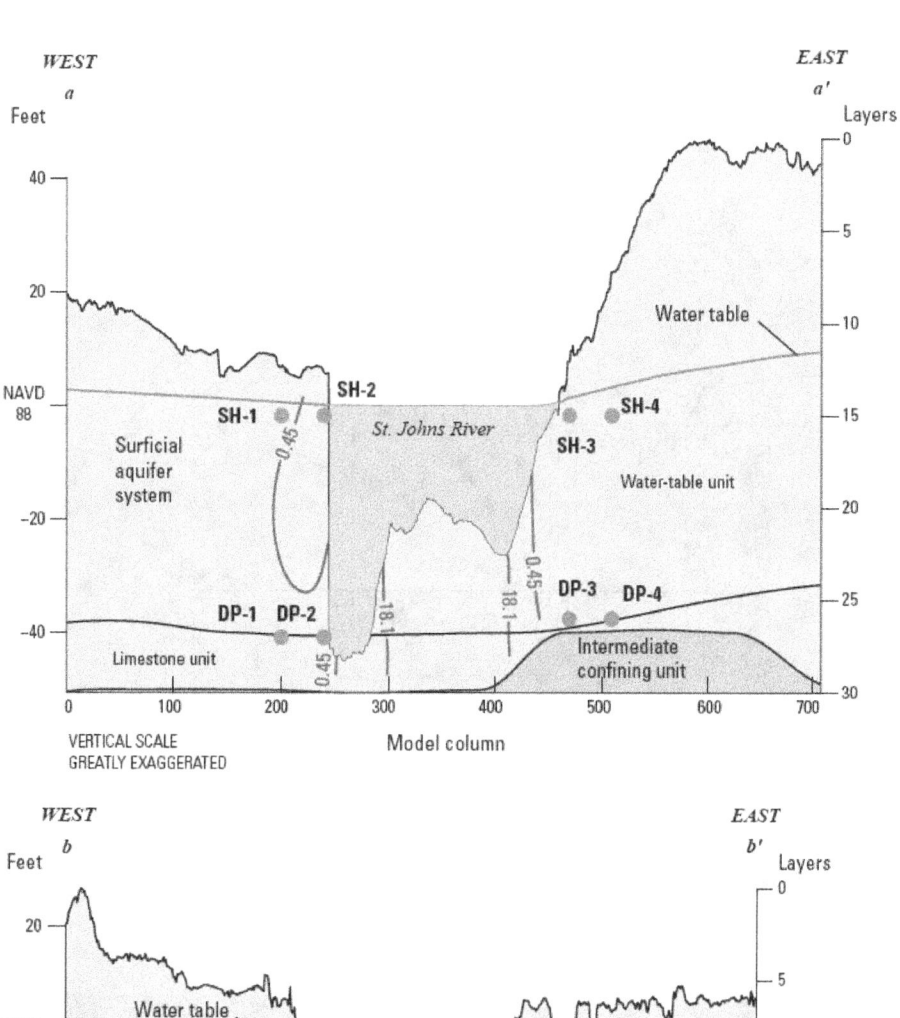

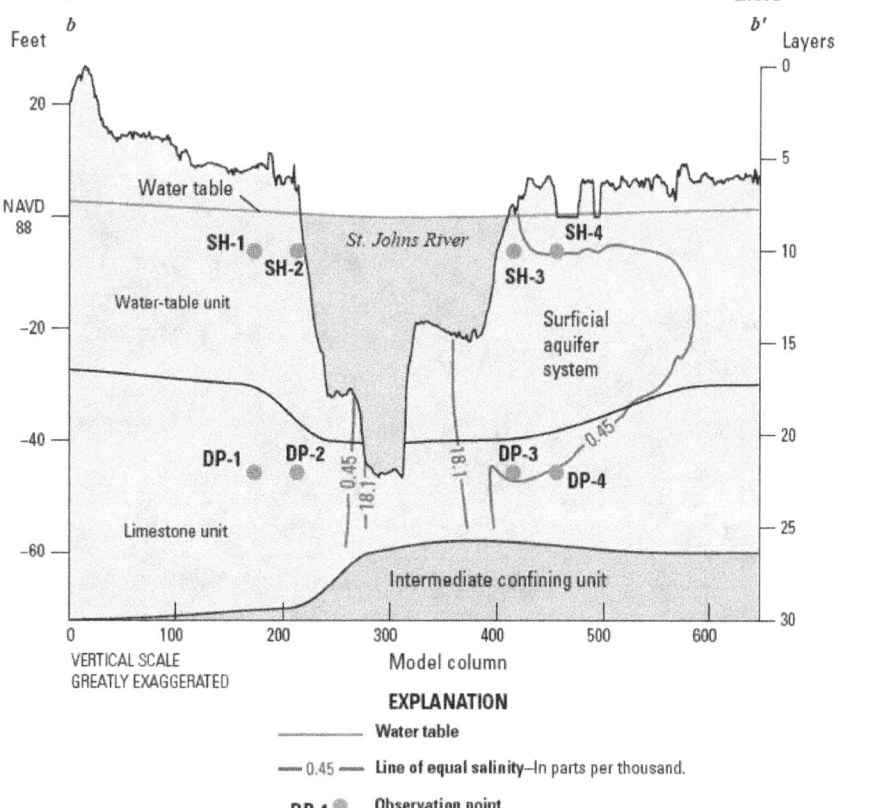

EXPLANATION

——————— Water table

—— 0.45 —— Line of equal salinity—In parts per thousand.

DP-1 ● Observation point

Figure 13. Undredged salinity contours (0.45 and 18.1 ppt) for base-case scenario; cross-sectional models *a–a'* and *b–b'*.

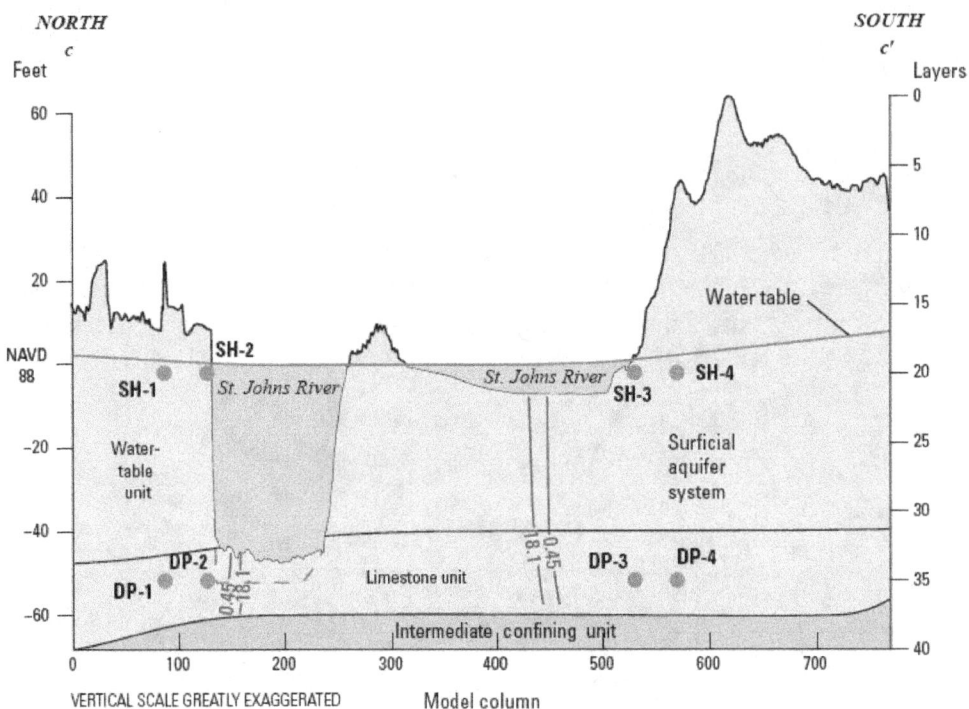

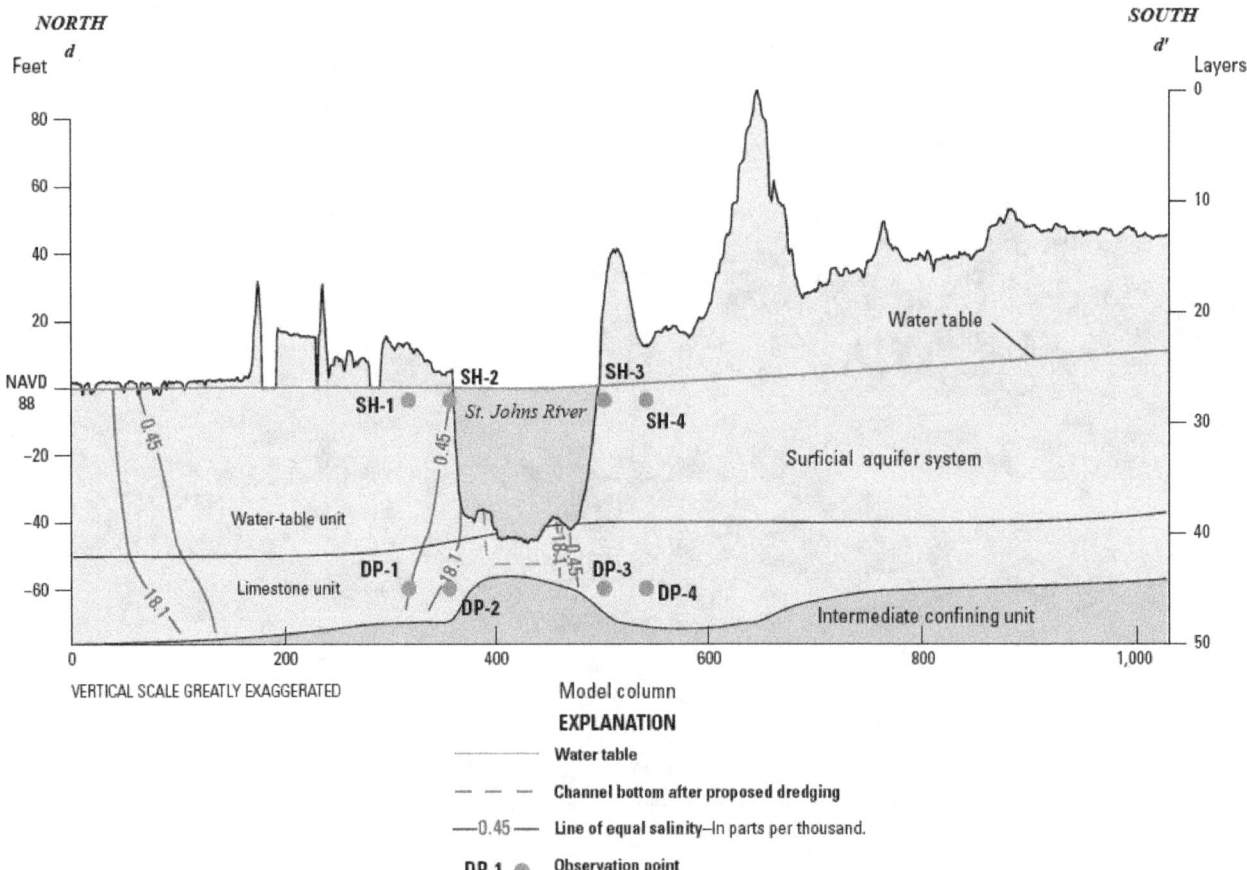

Figure 14.   Undredged salinity contours (0.45 and 18.1 ppt) for base-case scenario; cross-sectional models *c–c'* and *d–d'*.

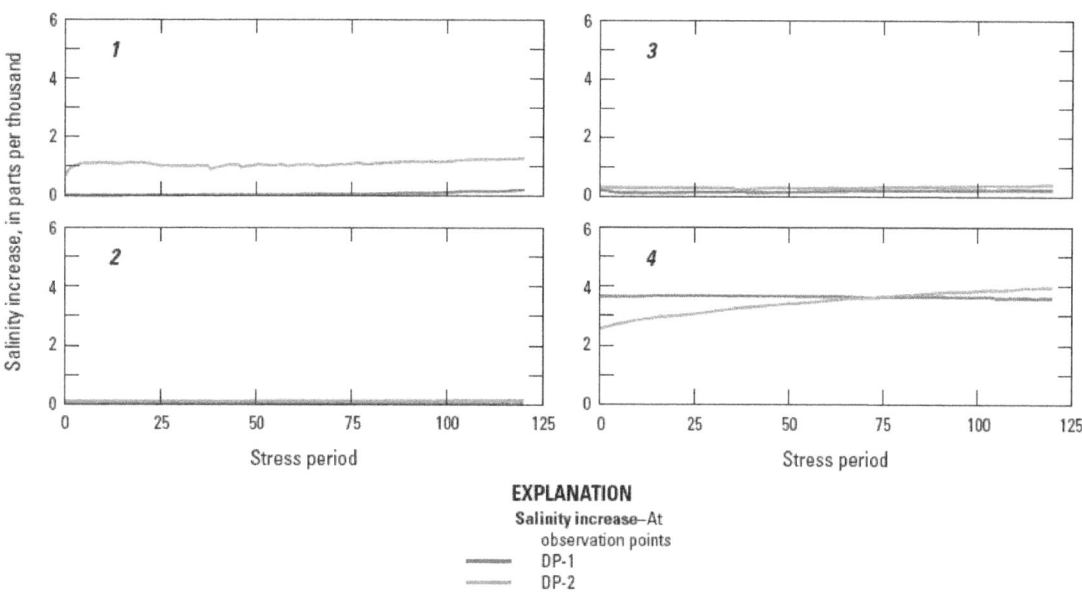

Figure 15. Computed salinity increases at DP-1 and DP-2 for each daily stress period at model cross section *d–d'* (fig. 8) for 121-day simulations. Simulation pair numbers are listed in the upper left-hand corner of each plot; names and descriptions are given in table 2. Relative locations of observation points are shown in figures 13 and 14.

Table 4. Maximum computed salinity increases at observation points caused by proposed dredging for extended 363-day simulations.

[ppt, parts per thousand; SH, shallow observation well; DP, deep observation well; N/A, not applicable; --, value less than 0.05 parts per thousand]

Cross section	Scenario number (fig. 12)	Scenario name	Maximum computed salinity increases, in ppt							
			SH-1	SH-2	SH-3	SH-4	DP-1	DP-2	DP-3	DP-4
d–d'	1	Simple system	--	0.1	--	--	0.9	1.7	--	--
	2	Semiconfined system[1]	N/A	N/A	N/A	N/A	N/A	N/A	N/A	N/A
	3	Preferential flow system[1]	N/A	N/A	N/A	N/A	N/A	N/A	N/A	N/A
	4	Complex system	--	--	--	--	3.7	5.3	--	--

[1]Simulation pair not run with extended transient simulation period.

Extended Simulations

Selected simulation pairs were performed using an extended, 363-day transient simulation period (described earlier) at model cross section *d–d'* (table 4, fig. 16). The purpose of these simulations was to assess whether maximum increases in salinity occurred beyond the end of the 121-day simulation period because of lags in system response to transient stresses. Extended simulations were not run for simulation pairs because results from the 121-day simulations indicated the maximum increase in salinity had been reached. Results from the extended simulations indicate that peak salinity differences generally were reached within about 160 to 190 days (fig. 16). All salinity increases for the extended 363-day simulations were greater than or equal to corresponding salinity increases for the 121-day simulations (tables 3 and 4). The simulation using the complex system (simulation pair 4) showed steady salinity increases at observation point DP-2 throughout the modeled period, and although it is not clear whether dynamic equilibrium was reached by the end of 363 days, the rate of change at the end of this simulation was less than the initial rates of change (fig. 16).

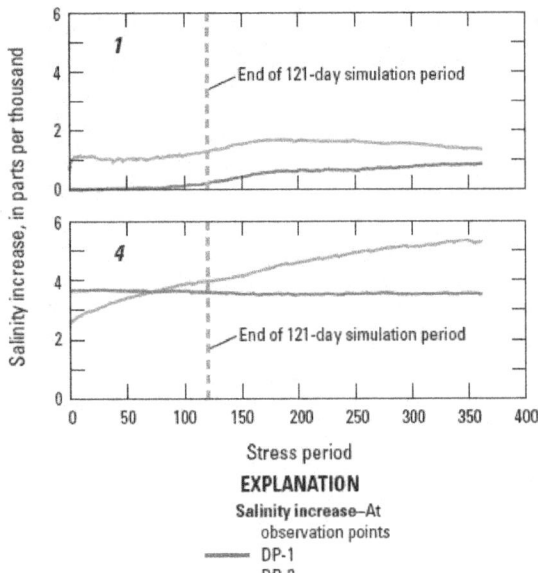

Figure 16. Computed salinity increases at DP-1 and DP-2 for each daily stress period at model cross section *d–d'* (fig. 8) for extended 363-day simulations. Simulation pair numbers are listed in the upper left-hand corner of each plot; names and descriptions are given in table 2. Relative locations of observation points are shown in figures 13 and 14.

Potential Effects of Channel Deepening on Saltwater Intrusion

Solutes are transported through porous aquifer materials by advective flow and hydrodynamic dispersion. Solutes carried by advection through a system travel at approximately the average linear velocity of the water. Advective flow is density dependent, meaning that water can seemingly flow upgradient because of the force of gravity. Complex advective flow patterns can develop within a variable-density groundwater system as denser saltwater flows from the river into the aquifer beneath the freshwater portion of the system discharging from the aquifer into the river. Heterogeneities in porous aquifer materials cause some solute particles to move at speeds faster or slower than the average linear velocity of the water in which they are dissolved (Anderson and Woessner, 2002). This phenomenon is known as hydrodynamic dispersion and mainly affects the distribution of solutes around the edge of a solute plume. This process causes the leading edge of a saltwater wedge to dilute as the front moves through the medium.

Simulation results for modeled conditions indicate that dredging will have little or no effect on groundwater salinity at model cross sections *a–a'*, *b–b'*, or *c–c'* (table 3) although dredging activities are expected to alter river salinity along the entire navigation channel, even in upstream locations that are not a part of the current dredging plans (Steve Bratos,

U.S. Army Corps of Engineers, written commun., 2012). Results of these model simulations are not discussed further.

Simulation results for modeled conditions indicate that dredging will have little to no effect on groundwater salinity in the water-table unit of the surficial aquifer system at model cross section *d–d'* and that increases in groundwater salinity generally would be less than 0.2 ppt, but could be as much as 4.0 ppt (table 3) in the limestone portion of the surficial aquifer system. All increases were indicated north of the river. The presence of semiconfining beds and high-conductivity preferential flow zones exacerbated increases in simulated aquifer salinity in this area. Low permeability beds lead to reductions in freshwater recharge to the limestone unit that may allow saline water from the river as well as areas beneath the upgradient salt marshes to flow through the highly conductive portions of the aquifer more easily. The effects of increased and decreased groundwater withdrawal rates were tested by multiplying the values in the WEL package by 100 and 0. Results indicated that neither scenario appreciably affected the groundwater-flow system or the salinity recorded at observation points.

Salinity changes indicated by model simulations are the result of two outcomes of dredging on the system: (1) altered shape and depth of the navigation channel and (2) increased river salinity. Results from the EFDC hydrodynamic model simulations indicate changes in river salinity of −1.5 to +3.2 ppt, averaging +1.7 ppt. An analysis of the relative importance of the outcomes of dredging showed that channel morphology generally has a greater effect on salinity variation in the limestone unit than does river salinity. Channel morphology and river salinity had about the same effect on salinity variation in the water-table unit.

Of the hydrogeologic realizations tested, those including only laterally extensive low-permeability beds seem most plausible. These beds range from 5 to 40 feet in thickness and from high to very low permeability and are known to be discontinuous in the study area, but this does not preclude the presence of laterally extensive beds similar to those modeled. Laterally extensive networks of highly transmissive fractures in the limestone unit may occur in the study area, but the likelihood that they intersect the proposed navigation channel bottom is unknown. The unit is cavernous near the study area (Fairchild, 1972; Causey and Phelps, 1978); however, fracture zones have not been mapped, and the unit is missing or discontinuous in some places along the coast (Phelps, 1994).

Sensitivity Analysis

The sensitivity of the model to changes in hydraulic conductivity and boundary heads was evaluated to determine their effect on simulated aquifer salinity differences for the simple system. Hydraulic conductivity and head parameters were altered for each sensitivity test, and resulting salinity differences were compared to the differences computed for the unperturbed case. Sensitivity is calculated as the difference between dredged and undredged salinity differences for each sensitivity test and the dredged and undredged salinity differences for the unperturbed case using the formula

$$S_x = [C' - C]_x - [C' - C]_u \qquad (4)$$

where

S is sensitivity of test x, relative to the unperturbed case, u;

C' is dredged salinity concentration; and

C is undredged salinity concentration.

Hydraulic conductivity parameters exert control on the potential for saltwater intrusion into the surficial aquifer system through an inverse relation to heads and, by extension, head gradients between the aquifer and river. Decreases in hydraulic conductivity increase heads and increases in hydraulic conductivity decrease heads, all other variables remaining unchanged. As heads in the aquifer change, so does the head gradient between the aquifer and river. Sensitivity test results indicated little appreciable sensitivity to changes in hydraulic conductivity at model cross sections a–a', b–b', and c–c' (table 5). Greatest sensitivity, +2.3 ppt, was indicated at observation point DP–2 of cross-sectional model d–d' when the horizontal hydraulic conductivity of limestone was decreased. The increase in salinity differences may seem counterintuitive because decreases in horizontal hydraulic conductivity, which increase the head gradient from the aquifer to the river, should reduce saltwater intrusion. Results show, however, that although differences between dredged and undredged conditions were in fact larger than for the unperturbed case, the overall magnitude of salinity decreased when horizontal hydraulic conductivity was reduced.

Boundary head elevations directly control the magnitude and direction of the head gradient between the aquifer and river and thus, the potential for saltwater intrusion. Lowered sufficiently, boundary heads could alter flow in an aquifer that discharges water to the river such that the flow direction is reversed and water flows from the river into the aquifer. Testing the effects of boundary head adjustments on simulation results was necessary to quantify the effects of the computed water-table elevations and boundary heads used in the absence of an adequate number of available field measurements. Boundary heads were decreased by multiplying the calculated water-table elevation from which they are derived by 0.8, 0.5, 0.2, 0.1, and 0.0. Lowering boundary heads increased the overall salinity of the aquifer, but generally resulted in relatively small changes in salinity differences between dredged and undredged conditions, compared with the unperturbed case (table 6). Greatest sensitivity, +3.3 ppt, was at observation point DP–2 of cross-sectional model c–c' with a boundary head elevation multiplier of 0.2.

Model Limitations

Numerical groundwater-flow models of natural systems are based on limited physical parameter and observation datasets. As a result, simplifications are incorporated into those models to overcome sparse datasets, which in turn limit the ability of the models to predict actual hydraulic conditions in a groundwater-flow system over time. Accuracy and predictive capability of groundwater-flow models developed for this study are mainly affected by spatial and temporal discretization and by uncertainty in system conceptualization and parameters, including boundary conditions and aquifer properties.

The three-dimensional variable-density groundwater-flow system was approximated in this study using two-dimensional models. Gravitational instabilities create complex advective flow patterns in variable-density groundwater-flow systems, and the resulting saltwater plumes are three-dimensional phenomena. It is assumed, however, that these cross-sectional models reasonably approximate the intrusion of relatively high salinity river water into the surficial aquifer. Care was taken when orienting the cross-sectional models used in this study to ensure they were approximately parallel to estimated groundwater flow lines in order to limit horizontal flow into and out of the plane of the model profiles. At or near sharp bends in the river, however, lateral groundwater flow lines are likely to be nonparallel, either converging or diverging, thereby introducing some error to model estimates. Groundwater withdrawals from wells can create horizontal flow into and out of the plane of the cross-sectional models. Withdrawals from the surficial aquifer system are minor, however, and thought to be relatively dispersed throughout the modeled areas; therefore, these withdrawals are not likely to induce substantial near-well lateral flow relative to the larger natural flow system.

The abbreviated simulation period of this study does not allow for the analysis of the effects of seasonal and inter-annual climate variability. The study area experiences roughly half of the yearly rainfall from June through September during high-intensity rainfall events such as thunderstorms, with the remainder distributed relatively evenly from October through May. As a result, the water table is generally highest at the end of the wet season, around September or October, and falls throughout the dry season to its lowest level, typically in May. The surficial aquifer system would potentially be most vulnerable, if at all, to saltwater intrusion when the minimum level is reached. The simulation period for this study ends in March, and the full effects of the dry season on head gradients in the aquifer are therefore not captured. The simulation period was exceptionally dry, however, and provides adequate conditions to test the effects of head gradients on saltwater intrusion in all but the most extreme cases. The simulation period was selected to correspond with the simulation period used by the EFDC model because results from that model were used in the groundwater-flow model presented herein.

Water-level data in the study area are available at neither the spatial nor the temporal scales needed to produce a map of the water table with sufficient detail for this study. Instead,

Table 5. Model sensitivity to changes in hydraulic conductivity parameters.

[SH, shallow observation point; DP, deep observation point; max., maximum; min., minimum; avg., average; hk, horizontal hydraulic conductivity; vk, vertical hydraulic conductivity; --, absolute value of change is less than 0.05 parts per thousand]

Cross section	Aquifer unit	Aquifer property	Multiplication factor	SH-1 Max.	SH-1 Min.	SH-1 Avg.	SH-2 Max.	SH-2 Min.	SH-2 Avg.	SH-3 Max.	SH-3 Min.	SH-3 Avg.	SH-4 Max.	SH-4 Min.	SH-4 Avg.	DP-1 Max.	DP-1 Min.	DP-1 Avg.	DP-2 Max.	DP-2 Min.	DP-2 Avg.	DP-3 Max.	DP-3 Min.	DP-3 Avg.	DP-4 Max.	DP-4 Min.	DP-4 Avg.
a–a'	Limestone	hk	0.2	--	--	--	--	--	--	--	--	--	--	--	--	--	--	--	--	--	--	--	--	--	--	--	--
		hk	5	--	--	--	--	--	--	--	--	--	--	--	--	--	--	--	--	--	--	--	--	--	--	--	--
		vk	5	--	--	--	--	--	--	--	--	--	--	--	--	--	--	--	--	--	--	--	--	--	--	--	--
	Water table	hk	0.2	--	--	--	--	--	--	--	--	--	--	--	--	--	--	--	--	--	--	--	--	--	--	--	--
		hk	10	--	--	--	--	--	--	--	--	--	--	--	--	--	--	--	--	--	--	--	--	--	--	--	--
		vk	5	--	--	--	--	--	--	--	--	--	--	--	--	--	--	--	--	--	--	--	--	--	--	--	--
b–b'	Limestone	hk	0.2	--	--	--	--	--	--	--	--	--	--	--	--	--	--	--	--	--	--	--	--	--	--	--	--
		hk	5	--	--	--	--	--	--	--	--	--	--	--	--	--	--	--	--	--	--	--	--	--	--	--	--
		vk	5	--	--	--	--	--	--	--	--	--	--	--	--	--	--	--	--	--	--	--	--	--	--	--	--
	Water table	hk	0.2	--	--	--	--	--	--	--	--	--	--	--	--	--	--	--	--	--	--	--	--	--	--	--	--
		hk	10	--	--	--	--	--	--	--	--	--	--	--	--	--	--	--	--	--	--	--	--	--	--	--	--
		vk	5	--	--	--	--	--	--	--	--	--	--	--	--	--	--	--	--	--	--	--	--	--	--	--	--
c–c'	Limestone	hk	0.2	--	--	--	--	--	--	--	--	--	--	--	--	--	--	--	--	--	--	--	--	--	--	--	--
		hk	5	--	--	--	--	--	--	--	--	--	--	--	--	--	--	--	--	--	--	--	--	--	--	--	--
		vk	5	--	--	--	--	--	--	--	--	--	--	--	--	--	--	--	--	--	--	--	--	--	--	--	--
	Water table	hk	0.2	--	--	--	--	--	--	--	--	--	--	--	--	--	--	--	--	--	--	--	--	--	--	--	--
		hk	10	--	--	--	--	--	--	--	--	--	--	--	--	--	--	--	0.2	--	0.1	--	--	--	--	--	--
		vk	5	--	--	--	--	--	--	--	--	--	--	--	--	--	--	--	--	--	--	--	--	--	--	--	--
d–d'	Limestone	hk	0.2	--	--	--	--	--	--	--	--	--	--	--	--	0.1	0.1	0.1	2.3	1.7	1.8	--	--	--	--	--	--
		hk	5	--	--	--	--	-0.1	--	--	--	--	--	--	--	0.7	0.3	0.5	1.2	--	0.7	--	--	--	--	--	--
		vk	5	--	--	--	--	--	--	--	--	--	--	--	--	0.2	--	0.1	1.1	0.2	0.8	--	--	--	--	--	--
	Water table	hk	0.2	--	--	--	--	--	--	--	--	--	--	--	--	--	--	--	0.1	--	--	--	--	--	--	--	--
		hk	10	-0.1	-0.3	-0.2	--	-0.2	-0.1	--	--	--	--	--	--	0.8	0.8	0.8	0.9	0.7	0.8	--	--	--	--	--	--
		vk	5	--	--	--	--	--	--	--	--	--	--	--	--	0.1	--	--	0.7	0.5	0.6	--	--	--	--	--	--

Table 6. Model sensitivity to changes in boundary head elevation.

[SH, shallow observation well; DP, deep observation well; max., maximum; min., minimum; avg., average; --, absolute value of change is less than 0.05 parts per thousand]

Cross section	Boundary head elevation multiplication factor	SH-1			SH-2			SH-3			SH-4			DP-1			DP-2			DP-3			DP-4		
		Max.	Min.	Avg.	Max.	Min.	Avg.	Max.	Min.	Avg.	Max.	Min.	Avg.	Max.	Min.	Avg.	Max.	Min.	Avg.	Max.	Min.	Avg.	Max.	Min.	Avg.
a–a'	0.0	--	--	--	0.1	--	--	--	--	--	--	--	--	--	--	--	--	--	--	--	--	--	--	--	--
	0.1	--	--	--	0.1	--	--	--	--	--	--	--	--	--	--	--	--	--	--	--	--	--	--	--	--
	0.2	--	--	--	--	--	--	--	--	--	--	--	--	--	--	--	0.1	--	--	--	--	--	--	--	--
	0.5	--	--	--	--	--	--	--	--	--	--	--	--	--	--	--	--	--	--	--	--	--	--	--	--
	0.8	--	--	--	--	--	--	--	--	--	--	--	--	--	--	--	--	--	--	--	--	--	--	--	--
b–b'	0.0	--	--	--	0.1	--	--	--	--	--	--	--	--	--	--	--	--	--	--	--	--	--	--	--	--
	0.1	--	--	--	--	--	--	--	--	--	--	--	--	--	--	--	--	--	--	--	--	--	--	--	--
	0.2	--	--	--	--	--	--	--	--	--	--	--	--	--	--	--	--	--	--	--	--	--	--	--	--
	0.5	--	--	--	--	--	--	--	--	--	--	--	--	--	--	--	--	--	--	--	--	--	--	--	--
	0.8	--	--	--	--	--	--	--	--	--	--	--	--	--	--	--	--	--	--	--	--	--	--	--	--
c–c'	0.0	--	--	--	0.4	0.1	0.3	0.1	--	--	--	--	--	--	--	--	1.7	--	1.1	--	--	--	--	--	--
	0.1	--	--	--	0.1	--	0.1	--	--	--	--	--	--	0.1	--	0.1	1.8	0.1	1.2	--	--	--	--	--	--
	0.2	--	--	--	--	--	--	--	--	--	--	--	--	--	--	--	3.3	1.0	2.3	--	--	--	--	--	--
	0.5	--	--	--	--	--	--	--	--	--	--	--	--	0.1	--	--	0.1	--	--	--	--	--	--	--	--
	0.8	--	--	--	--	--	--	--	--	--	--	--	--	--	--	--	--	--	--	--	--	--	--	--	--
d–d'	0.0	--	--	--	--	--	--	--	--	--	--	--	--	--	--	--	--	--	--	1.0	0.6	0.8	-0.1	-0.1	-0.1
	0.1	--	--	--	--	--	--	--	--	--	--	--	--	--	-0.1	--	--	-0.3	-0.2	0.5	--	0.1	--	--	--
	0.2	--	--	--	--	--	--	--	--	--	--	--	--	--	--	--	--	-0.3	-0.2	--	--	--	--	--	--
	0.5	--	--	--	--	--	--	--	--	--	--	--	--	0.1	0.1	0.1	0.1	-0.3	-0.2	--	--	--	--	--	--
	0.8	--	--	--	--	--	--	--	--	--	--	--	--	0.1	0.1	0.1	0.1	-0.2	-0.2	--	--	--	--	--	--

a multiple-linear-regression equation was used to estimate the mean annual water-table elevation and provide boundary conditions for the steady-state models. This method is unable to predict the potential formation of perched groundwater mounds on top of clayey sand lenses in the water-table unit of the surficial aquifer system. It is likely that this method for estimating mean annual water levels overestimated the actual elevation of the water-table surface during the ongoing drought in the simulation period. Computed water levels were compared to observed, daily maximum water levels recorded in two observation wells from December 1, 1998, to March 31, 1999, to assess the accuracy of the method. Daily maximum water levels recorded at well DS–520 SJRWMD Observation Well at Jacksonville, FL (USGS site number 301710081323601) (fig. 1, table 1) ranged from 40.73 to 41.55 feet above NAVD 88 (U.S. Geological Survey, 2013b); the computed water level at this location was 47.83 feet above NAVD 88. Daily maximum water levels recorded at well DS–522 Fort Caroline National Memorial Park (USGS site number 302301081295001) (fig. 1, table 1) ranged from 7.65 to 8.75 feet above NAVD 88 (U.S. Geological Survey, 2013b); the computed water level at this location was 10.20 feet above NAVD 88.

Head-dependent flux boundaries were specified using the GHB package to simulate fluxes into and out of the cell faces along the ends of each cross-sectional model instead of no-flow boundaries because physical and (or) hydrologic boundaries were not close enough to be reasonably used. The boundaries of an ideal groundwater model will extend to the physical boundaries of the system being simulated, but this is not feasible in many cases. The use of head-dependent flux boundaries using the GHB package was preferred to specified head boundaries, even though boundary heads did not vary with time because fluxes through cells specified with the GHB package are calculated using a conductance term, C (eq. 2), that acts to retard flow between the boundary cell and the model according to Darcian principles.

Few data describing aquifer properties were available for the surficial aquifer system in Duval County. Published hydraulic-test data for the aquifer are limited, yielding a few transmissivity and horizontal conductivity values for the water-table and limestone units. No hydraulic conductivity data were available for the semiconfining unit. Uniform parameter values were used to represent the hydraulic properties of each discrete layer or feature of the surficial aquifer system. Aquifer properties were assigned to the models using a combination of field data and published values typical of the materials that compose the aquifer.

Sensitivity analyses of dispersivity and porosity were not conducted. These parameters control the distribution and salinity of a saltwater plume as it moves through an aquifer. Increasing dispersivity and porosity values increase the areal distribution of saltwater, but decrease salinity (diffuse plume); decreasing the dispersivity value decreases areal distribution, but increases salinity (sharp-interface plume). Dispersivity is a scale-dependent parameter in that it increases with distance travelled by the plume (Anderson and Woessner, 2002).

Dispersivity values for the models presented herein were taken from models developed by Langevin (2001) for limestone aquifers underlying Biscayne Bay in southern Florida. The scales of the Biscayne Bay models bracket those of the models for this study and are of the same order of magnitude. It was therefore assumed that the average of the dispersivity values used in the Biscayne Bay models was appropriate for the models used in this study. The porosity values used for the models described herein are equal to those of specific yield and represent the arithmetic mean of the range of values in Anderson and Woessner (2002, table 3.5) for the equivalent aquifer media type.

Calibration of the models was not possible because the requisite water-level and salinity data were not available. The calibration process is used to more closely align model parameter values (hydraulic conductivity, specific yield, dispersivity, and so forth) with those of the system being simulated by comparing model results with observed data. The hypothetical nature of the scenarios tested in this study allows for the use of reasonable parameter sets with uncalibrated models to predict system responses to those scenarios. The ultimate goal of this study was not to simulate conditions as they currently exist, but to determine whether proposed dredging will increase salinity in the surficial aquifer system under "worst case" circumstances.

Accurate metered data for domestic pumping were not available, so county-wide estimates were used to quantify water use within the study area. An effort was made to reasonably distribute water use spatially, but pumpage was applied at a constant rate throughout the transient simulation period. Preliminary testing showed that models were insensitive to groundwater withdrawals. Model pumping rates were varied from estimated values (multiplied by 0 and 100) with little to no effect on the simulated flow system, suggesting that any uncertainty in pumping rate does not affect model reliability. This result was expected, given that groundwater withdrawals are a relatively small part of the water budget—nearly two orders of magnitude smaller than precipitation or ET.

Net recharge applied to the surficial aquifer system was calculated from the best available sources of precipitation and ET data. Uncertainty in these datasets exists, however, particularly when applying local corrections for runoff in the case of precipitation and crop coefficients in the case of ET. Variations in land-cover permeability, the degree of soil saturation, and slope can affect actual runoff, whereas variations in vegetation type, soil moisture availability, and shade can affect the actual ET. The location, total number, and daily throughput of septic systems in the study area are poorly known, and each affects the calculation of return flow to the surficial aquifer from septic systems. A sensitivity analysis of net recharge was not performed explicitly; decreased heads used in the sensitivity analysis of boundary heads serve to approximate the conditions expected to occur if net recharge was decreased. Results from the boundary-head sensitivity analysis therefore act as a proxy by which the effects of reductions in net recharge on aquifer salinity can be examined.

Water percolation through the unsaturated zone is an important factor to account for when computing net recharge, but it was not accounted for in this study. Instead, infiltrating water at land surface was assumed to reach the water table instantaneously without any change in magnitude. Actual water-table recharge for a given model stress period could be higher or lower than that prescribed in the model because of the storative and time-delaying effect of the unsaturated zone; nevertheless, such recharge would be unbiased in magnitude over an extended time period.

Summary

The U.S Army Corps of Engineers has proposed dredging a 13-mile reach of the navigation channel in Jacksonville, Florida, deepening it to depths between 50 and 54 feet below the North American Vertical Datum of 1988 (NAVD 88). The dredging operation will remove about 10 feet of sediments from the surficial aquifer system, including limestone in some locations. The limestone unit, which is in the lowermost part of the surficial aquifer system, supplies water to domestic wells in the Jacksonville area. Because of the density-driven hydrodynamics of the river, saline water from the Atlantic Ocean travels upstream as a saltwater "wedge" along the bottom of the channel, precisely where the limestone is most likely to be exposed by the proposed dredging.

Simulations were run with each of four cross-sectional, variable-density groundwater-flow models, developed using SEAWAT, to simulate hypothetical changes in salinity in the surficial aquifer system as a result of dredging. Simulated results for modeled conditions indicate that dredging will have little to no effect on salinity variations in areas upstream of currently proposed dredging activities. Results also indicate little to no effect in any part of the surficial aquifer system along model cross section $c–c'$ or in the water-table unit along model cross section $d–d'$. Increases of up to 4.0 parts per thousand (ppt) were indicated by the model incorporating the complex hydrogeologic realization along model cross section $d–d'$. Simulated increases in salinity greater than 0.2 ppt in this area were generally limited to portions of the limestone unit within about 75 feet of the channel on the north side of the river. Extended 363-day transient simulations evaluated whether peak salinity increases from selected 121-day transient simulations were affected by a lag in response of the system to transient stresses, such as net recharge, river stage, river salinity, and so forth. Results from these extended simulations indicated that all peak salinities for the 363-day simulations were greater than or equal to those for the 121-day simulations. Differences in peak salinities averaged 0.4 ppt and did not exceed 1.6 ppt.

The potential for saltwater to move from the river channel to the surficial aquifer system is limited, but may be present in areas where the head gradient from the aquifer to the river is small or negative and the salinity of the river is sufficient to induce density-driven advective flow into the aquifer. In some areas, simulated increases in salinity were exacerbated by the presence of laterally extensive semiconfining beds in combination with a high-conductivity preferential flow zone in the limestone unit of the surficial aquifer system and an upgradient source of saline water, such as beneath the salt marshes near Fanning Island. The volume of groundwater pumped in these areas is estimated to be low; therefore, saltwater intrusion will not substantially affect regional water supply, although users of the surficial aquifer system east of Dames Point along the northern shore of the river could be affected. Proposed dredging operations pose no risk to salinization of the Floridan aquifer system; in the study area, the intermediate confining unit ranges in thickness from more than 300 to about 500 feet and provides sufficient hydraulic separation between the surficial and Floridan aquifer systems.

The cross-sectional models developed in this study do not necessarily simulate actual conditions. Instead, the models were used to examine the potential effects of deepening the navigation channel on saltwater intrusion in the surficial aquifer system under a range of plausible hypothetical conditions. Based on simulation results of such conditions, the risk of dredging-induced saltwater intrusion affecting the water supply is estimated to be low. The largest simulated increases in groundwater salinity were mainly in areas where there is little demand for groundwater from the surficial aquifer system. Groundwater levels and water quality would need to be monitored, particularly in the limestone unit along the northern periphery of the river channel near model cross section $d–d'$, to determine if any changes in salinity occur within the aquifer after the channel has been dredged. Such data would also aid the calibration of a three-dimensional model of the system that could more accurately assess the impacts of dredging on water quality in the surficial aquifer system.

Selected References

Anderson, M.P., and Woessner, W.W., 2002, Applied groundwater modeling—Simulation of advective flow and transport: San Diego, Calif., Academic Press, 381 p.

Anderson, Warren, and Goolsby, D.A., 1973, Flow and chemical characteristics of the St. Johns River at Jacksonville, Florida: Florida Bureau of Geology Information Circular no. 82, 57 p.

Boniol, Don, Williams, Marvin, and Munch, Douglas, 1993, Mapping recharge to the Floridan aquifer using a geographic information system: St. Johns River Water Management District Technical Publication SJ93–5, 41 p.

Causey, L.V., and Phelps, G.G., 1978, Availability and quality of water from shallow aquifers in Duval County, Florida: U.S. Geological Survey Water-Resources Investigations Report 78–92, 36 p.

Daly, Christopher, and Gibson, W., 2002, 103-Year high-resolution precipitation climate data set for the conterminous United States: Corvallis, Ore., The PRISM Climate Group, accessed April 26, 2010, at *http://prism.oregonstate.edu.*

Daly, Christopher, Gibson, W., Doggett, M., and Smith, J., 2011, Near-real-time monthly high-resolution precipitation climate data set for the conterminous United States: Corvallis, Ore., The PRISM Climate Group, accessed April 26, 2010, at *http://prism.oregonstate.edu.*

Dames and Moore, 1987, Site proposal superconducting super collider: Geology and Tunneling, v. 3: Consultant's report prepared for the State of Florida, variously paged.

Davis, J.H., Planert, M., and Andrews, W.J., 1996, Simulation of ground-water flow at the U.S. Naval Air Station, Jacksonville, Florida, with an evaluation of changes to ground-water movement caused by proposed remedial designs at Operable Unit 1: U.S. Geological Survey Open-File Report 96–597, 47 p.

Dexco, Inc., 2009, Jacksonville Harbor resistivity survey: Consultant's report prepared for the U.S. Army Corps of Engineers, variously paged.

Fairchild, R.W., 1972, The shallow-aquifer system in Duval County, Florida: Florida Bureau of Geology Report of Investigations 59, 50 p.

Faye, R.E., and Mayer, G.C., 1990, Ground-water flow and stream-aquifer relations in the northern Coastal Plain of Georgia and adjacent parts of Alabama and South Carolina: U.S. Geological Survey Water-Resources Investigations Report 88–4143, 83 p.

Fetter, C.W., 2001, Applied hydrogeology: Upper Saddle River, N.J., Prentice-Hall, 598 p.

Franks, B.J., 1980, The surficial aquifer system at the U.S. Naval Station near Mayport, Florida: U.S. Geological Survey Open-File Report 80–795, 13 p.

Gesch, D.B., 2007, The National Elevation Dataset, *in* Maune, D.F., ed., Digital elevation model technologies and applications—The DEM users manual (2d ed): Bethesda, Md., American Society for Photogrammetry and Remote Sensing, p. 99–118.

Gesch, D.B., Oimoen, M., Greenlee, S., Nelson, C., Steuck, M., and Tyler, D., 2002, The National Elevation Dataset: Photogrammetric Engineering and Remote Sensing, v. 68, no. 1, p. 5–11.

Guo, Weixing, and Langevin, C.D., 2002, User's Guide to SEAWAT—A computer program for simulation of three-dimensional variable-density ground-water flow: U.S. Geological Survey Techniques of Water-Resources Investigations, book 6, chap. A7, 77 p.

Halford, K.J., 1998a, Ground-water flow in the surficial aquifer system and potential movement of contaminants from selected waste-disposal sites at Cecil Field Naval Air Station, Jacksonville, Florida: U.S. Geological Survey Water-Resources Investigations Report 97–4278, 68 p.

Halford, K.J., 1998b, Ground-water flow in the surficial aquifer system and potential movement of contaminants from selected waste-disposal sites at Naval Station Mayport, Florida: U.S. Geological Survey Water-Resources Investigations Report 97–4262, 104 p.

Hall, Pamela, and Clancy, S.J., 2009, The Florida statewide inventory of onsite sewage treatment and disposal systems (OSTDS): Final report to the State of Florida Department of Health, Division of Environmental Health, Bureau of Onsite Sewage Programs, 157 p.

Hamrick, J.M., 1992, A three-dimensional environmental fluid dynamics computer code—Theoretical and computational aspects: The College of William and Mary, Virginia Institute of Marine Science Special Report 317, 63 p.

Harbaugh, A.W., and McDonald, M.G., 1996, User's documentation for MODFLOW–96, an update to the U.S. Geological Survey modular finite-difference ground-water flow model: U.S. Geological Survey Open-File Report 96–485, 56 p.

Harbaugh, A.W., Banta, E.R., Hill, M.C., and McDonald, M.G., 2000, MODFLOW–2000, the U.S. Geological Survey modular ground-water model—User guide to modularization concepts and the ground-water flow process: U.S. Geological Survey Open-File Report 00–92, 121 p.

Jacobs, Jennifer, Mecikalski, J., and Paech, S., 2008, Satellite-based solar radiation, net radiation, and potential and reference evapotranspiration estimates over Florida: Technical report to the U.S. Geological Survey (Also available at *http://fl.water.usgs.gov/et/publications/GOES_FinalReport.pdf.*)

Jia, Xinhua, Dukes, M.D., and Jacobs, J.M., 2009, Bahiagrass crop coefficients from eddy correlation measurements in central Florida: Irrigation Science, v. 28, no. 1, p. 5–15.

Knochenmus, L.A., 2006, Regional evaluation of the hydrogeologic framework, hydraulic properties, and chemical characteristics of the intermediate aquifer system underlying southern west-central Florida: U.S. Geological Survey Scientific Investigations Report 2006–5013, 40 p.

Konikow, L.F., 2011, The secret to successful solute-transport modeling: Ground Water, v. 49, no. 2, p. 144–159.

Langevin, C.D., 2001, Simulation of ground-water discharge to Biscayne Bay, southeastern Florida: U.S. Geological Survey Water-Resources Investigations Report 00–4251, 127 p.

Langevin, C.D., Shoemaker, W.B., and Guo, W., 2003, MODFLOW–2000, the U.S. Geological Survey modular ground-water model—Documentation of the SEA-WAT–2000 version with the Variable-Density Flow Process (VDF) and the Integrated MT3DMS Transport Process (IMT): U.S. Geological Survey Open-File Report 03–426, 43 p.

Langevin, C.D., Thorne, D.T., Dausman, A.M., Sukop, M.C., and Guo, W.W., 2008, SEAWAT Version 4—A computer program for simulation of multi-species solute and heat transport: U.S. Geological Survey Techniques and Methods, book 6, chap. A22, 39 p.

Leve, G.W., 1966, Ground-water in Duval and Nassau Counties, Florida: Florida Geological Survey Report of Investigations no. 43, 91 p.

Marella, R.L., 2004, Water withdrawals, use, discharge, and trends in Florida, 2000: U.S. Geological Survey Scientific Investigations Report 2004–5151, 136 p.

Marella, R.L., 2009, Water withdrawals, use, and trends in Florida, 2005: U.S. Geological Survey Scientific Investigations Report 2009–5125, 49 p.

McDonald, M.G., and Harbaugh, A.W., 1988, A modular three-dimensional finite-difference ground-water flow model: U.S. Geological Survey Techniques of Water Resources Investigations, book 6, chap. A1, 586 p.

Mecikalski, J.R., Sumner, D.M., Jacobs, J.M., Pathak, C.S., Paech, S.J., and Douglas, E.M., 2011, Use of visible geostationary operational meteorological satellite imagery in mapping reference and potential evapotranspiration over Florida, in Labedzki, Leszek, ed., Evapotranspiration: Vienna, Austria, InTech Publishers, 446 p. (Also available at http://www.intechopen.com/books/evapotranspiration/use-of-visible-geostationary-operational-meteorological-satellite-imagery-in-mapping-reference-and-p.)

Miller, J.A., 1986, Hydrogeologic framework of the Floridan aquifer system in Florida and in parts of Georgia, South Carolina, and Alabama: U.S. Geological Survey Professional Paper 1403–B, 91 p.

Morris, F.W., 1995, Volume 3 of the Lower St. Johns River Basin reconnaissance—Hydrodynamics and salinity of surface water: St. Johns River Water Management District Technical Publication SJ95–9, 362 p.

National Oceanic and Atmospheric Administration, 2013, Tidal datums, accessed June 21, 2013, at http://tidesandcurrents.noaa.gov/datum_options.html.

Phelps, G.G., 1994, Water resources of Duval County, Florida: U.S. Geological Survey Water-Resources Investigations Report 93–4130, 78 p.

Puri, H.S., 1957, Stratigraphy and zonation of the Ocala Group: Florida Geological Survey Bulletin 38, 258 p.

Puri, H.S., and Vernon, R.O., 1964, Summary of the geology of Florida and a guidebook of the classic exposures: Florida Geological Survey Special Publication 5, 312 p.

Reilly, T.E., 2001, System and boundary conceptualization in ground-water flow simulation: U.S. Geological Survey Techniques of Water Resources Investigations, book 3, chap. B8, 30 p.

Scott, T.M., 1988, The lithostratigraphy of the Hawthorn Group (Miocene) of Florida: Florida Geological Survey Bulletin 59, 148 p.

Sepúlveda, Nicasio, 2002, Simulation of ground-water flow in the intermediate and Floridan aquifer systems in Peninsular Florida: U.S. Geological Survey Water-Resources Investigations Report 02–4009, 130 p.

Spechler, R.M., 1994, Saltwater intrusion and quality of water in the Floridan aquifer system, northeastern Florida: U.S. Geological Survey Water-Resources Investigations Report 92–4174, 76 p.

Spechler, R.M., 1996, Detection and quality of previously undetermined Floridan aquifer system discharge to the St. Johns River, Jacksonville, to Green Cove Springs, northeastern Florida: U.S. Geological Survey Water-Resources Investigations Report 95–4257, 29 p.

Spechler, R.M., and Stone, R.B., Jr., 1983, Appraisal of the interconnection between the St. Johns River and the surficial aquifer system, east-central Duval County, Florida: U.S. Geological Survey Water-Resources Investigations Report 82–4109, 34 p.

St. Johns River Water Management District, 1994, St. Johns River, Florida, water quality feasibility study phase 1 interim report—Volume 1, Executive summary: Palatka, Florida, St. Johns River Water Management District Special Publication SJ94–SP12, 103 p.

St. Johns River Water Management District, 2004, SJRWMD land use and land cover (2004): Vector digital data available at ftp://secure.sjrwmd.com/disk6b/lcover_luse/luse2004/.

St. Johns River Water Management District, 2008, Lower St. Johns River salinity regime assessment—Effects of upstream flow reduction near Deland: Palatka, Florida, St. Johns River Water Management District Special Publication SJ2004–SP29, variously paged.

Tibbals, C.H., 1990, Hydrogeology of the Floridan aquifer system in east-central Florida: U.S. Geological Survey Professional Paper 1403–E, 98 p.

U.S. Army Corps of Engineers, 1986, St. Johns River Basin, Florida—Interim water quality management plan findings: U.S. Army Corps of Engineers, Jacksonville District, variously paged.

U.S. Army Corps of Engineers, 2009, Jacksonville Harbor Navigation Project: U.S. Army Corps of Engineers, Jacksonville District, 6 p.

U.S. Army Corps of Engineers, 2011, Jacksonville Harbor (Mile Point) navigational study, Duval County, Florida—Draft-integrated feasibility report and environmental assessment: U.S. Army Corps of Engineers, Jacksonville District, 128 p.

U.S. Census Bureau, 2012, 2010 Census data, accessed February 2012 at *http://www.census.gov/2010census/data/*.

U.S. Code of Federal Regulations, 2002, National secondary drinking water standards (7–1–02 ed.): Pt. 143, Sec. 3, p. 614.

U.S. Geological Survey, 2012, Florida Water Science Center evapotranspiration information and data (Statewide evapotranspiration data), accessed August 2012 at *http://fl.water.usgs.gov/et/*.

U.S. Geological Survey, 2013a, National Water Information System data available on the World Wide Web (USGS water-quality data for Florida), accessed May 2013 at *http://waterdata.usgs.gov/fl/nwis/qw*.

U.S. Geological Survey, 2013b, National Water Information System data available on the World Wide Web (USGS groundwater data for Florida), accessed June 2013 at *http://waterdata.usgs.gov/fl/nwis/gw*.

U.S. Geological Survey, 2013c, Water-resources data for the United States, Water Year 2012: U.S. Geological Survey Water-Data Report WDR-US-2012, site 02246500, accessed June 2013 at *http://wdr.water.usgs.gov/wy2012/pdfs/02246500.2012.pdf*.

Vernon, R.O., 1951, Geology of Citrus and Levy Counties, Florida: Florida Geological Survey Bulletin 33, 256 p.

Zheng, Chunmiao, and Wang, P.P., 1998, MT3DMS, A modular three-dimensional multispecies transport model for simulation of advection, dispersion, and chemical reactions of contaminants in groundwater systems: Vicksburg, Miss., Waterways Experiment Station, U.S. Army Corps of Engineers.

www.ingramcontent.com/pod-product-compliance
Lightning Source LLC
Chambersburg PA
CBHW081403170526
45166CB00010B/3191